CONTENTS

S. HRG. 103–911

S. 2101, THE COMPREHENSIVE ONE-CALL NOTIFICATION ACT OF 1994

HEARING

BEFORE THE

COMMITTEE ON COMMERCE, SCIENCE, AND TRANSPORTATION UNITED STATES SENATE

ONE HUNDRED THIRD CONGRESS

SECOND SESSION

AUGUST 3, 1994

Printed for the use of the Committee on Commerce, Science, and Transportation

U.S. GOVERNMENT PRINTING OFFICE

81–903 CC WASHINGTON : 1995

For sale by the U.S. Government Printing Office
Superintendent of Documents, Congressional Sales Office, Washington, DC 20402
ISBN 0-16-046587-7

COMMITTEE ON COMMERCE, SCIENCE, AND TRANSPORTATION

ERNEST F. HOLLINGS, South Carolina, *Chairman*

DANIEL K. INOUYE, Hawaii
WENDELL H. FORD, Kentucky
J. JAMES EXON, Nebraska
JOHN D. ROCKEFELLER IV, West Virginia
JOHN F. KERRY, Massachusetts
JOHN B. BREAUX, Louisiana
RICHARD H. BRYAN, Nevada
CHARLES S. ROBB, Virginia
BYRON L. DORGAN, North Dakota
HARLAN MATHEWS, Tennessee

JOHN C. DANFORTH, Missouri
BOB PACKWOOD, Oregon
LARRY PRESSLER, South Dakota
TED STEVENS, Alaska
JOHN McCAIN, Arizona
CONRAD BURNS, Montana
SLADE GORTON, Washington
TRENT LOTT, Mississippi
KAY BAILEY HUTCHISON, Texas

KEVIN G. CURTIN, *Chief Counsel and Staff Director*
JONATHAN CHAMBERS, *Republican Staff Director*

S. 2101, THE COMPREHENSIVE ONE-CALL NOTIFICATION ACT OF 1994

WEDNESDAY, AUGUST 3, 1994

U.S. SENATE,
COMMITTEE ON COMMERCE, SCIENCE, AND TRANSPORTATION,
Washington, DC.

The committee met, pursuant to notice, at 9:35 a.m. in room SR–253, Russell Senate Office Building, Hon. J. James Exon, presiding.

Staff members assigned to this hearing: Clyde J. Hart, Jr., senior counsel, William Clyburn, Jr., staff counsel, and Kevin M. Jospeh, professional staff member; and Gerri Lynn Hall, minority professional staff member.

OPENING STATEMENT OF SENATOR EXON

Senator EXON. The committee please will come to order.

I welcome my colleagues and distinguished experts to discuss legislation introduced by Senators Bradley and Lautenberg and Congressmen Pallone and Sharp, known as the Comprehensive One-Call Notification Act. Since Senators Lautenberg and Bradley have scheduling conflicts this morning, I will keep my opening remarks very brief so they may have time before their next scheduled appointment. I appreciate both of them being here.

This subcommittee has been very active in promoting pipeline safety, including support for one-call systems. One-call systems require individuals to make one call before digging into the ground to make certain that there are no buried pipeline cables or power lines which may be damaged by digging. We had a recent experience in my State of Nebraska with regard to laying of new cable lines, and I think, at the last report, there were some 24 separate incidents where other lines were severed, at great inconvenience to the public. I think considering this legislation is very worthwhile.

No Senate subcommittee has been more committed to pipeline safety and other safety issues than this one which I chair. In 1992, Congress approved comprehensive pipeline safety legislation which I introduced into the Senate. This bill included provisions to require the use of automatic shutoff valves, create civil penalties for those who cause damage while digging without first calling a one-call system, and require public pipeline companies to maintain pipeline locations and maps.

The legislation now before the committee builds on that record, attempts to improve the one-call system, and expand it to cover new areas. As a strong supporter of the one-call concept, I want to be assured about the cost and the effectiveness of the proposed bill to consumers and taxpayers alike. I will work closely with my colleagues from the Senate and House to craft workable, one-call legislation which can be enacted this year, and which, first and foremost, protects the public safety. Mr. Chairman, your comments, please.

OPENING STATEMENT OF SENATOR HOLLINGS

The CHAIRMAN. The Committee on Commerce, Science, and Transporation will receive testimony today on S. 2101, the Comprehensive One-Call Notification Act of 1994, introduced by Senator Bradley and cosponsored by Senator Lautenberg. This bill is designed to reduce the damage done to underground facilities by excavators with mechanized equipment.

The magnitude of this problem cannot be overstated. By one estimate, close to 20 million miles of pipe, cables, and wires make up this Nation's utility infrastructure. This number inevitably will grow in the coming years. The harm done to this underground complex by inadvertent excavating damage affects us all—whether it is the sudden and devastating explosions that have taken place lately in Indiana, New Jersey, Pennsylvania, Minnesota, and Wyoming, or the time-consuming and costly disruption to the public and business community that ensued when a cable was damaged here at Washington National Airport 6 weeks ago. A comprehensive one-call system is one step to prevent such harm to the country's citizens and economy.

But, we have to ensure that we tailor our solution to the precise problem. There are issues that must be resolved before we can declare that we have begun to meet the challenge before us. We need to examine and address issues such as how to develop an equitable fee structure for one-call operations, whether certain facilities or potential excavators should be exempt from the one-call system, and what penalties are appropriate for violations of the one-call system.

This hearing will help us to develop an appropriate response to such issues and to decide what action to take on S. 2101. I look forward to today's testimony.

Thank you, Mr. Chairman.

Senator EXON. Thank you, Mr. Chairman. I ask my colleagues, in an effort to accommodate everyone's schedule, to allow me to recognize Senators Lautenberg and Bradley first for opening statements. With that, following the customary rule, I recognize the senior Senator from the great State of New Jersey, Senator Bradley.

STATEMENT OF HON. BILL BRADLEY, U.S. SENATOR FROM NEW JERSEY

Senator BRADLEY. Thank you very much, Mr. Chairman. I am pleased to be here today to testify on behalf of comprehensive one-call legislation and to be here with my colleague Senator Lautenberg, and I appreciate you holding this hearing and your interest in this bill.

The bill you are considering will create new assurance that accidents involving pipelines and underground utilities will not occur. Every year, multiple fatalities and tens of millions of dollars' worth of damage occur simply because people dig where they should not. These third-party incidents are the single leading cause of accidents involving pipelines. According to the Department of Transportation, these accidents result in over one-half of the fatalities and one-half of the property damage caused by all pipeline failures. Legislation that I have introduced, the Comprehensive One-Call Notification Act, will create a mechanism to prevent the inadvertent injury and the potential tragedy.

Last March 23, just before midnight, an explosion ripped through the community of Durham Woods in Edison, NJ. Within minutes, eight apartment buildings were ablaze. Soon, they were gone, wiped out by a fireball that lit up the sky over hundreds of square miles. One life was lost, only one. Hundreds lost their homes. Many more were evacuated. The injuries were miraculously low, but who knows how many others still lie awake at night wondering whether it could happen again and fearing for their own future?

Reflecting on the accident today, it seems hard to fault anyone for their response to the tragedy. The community pulled together and helped out those in need. Food, emergency shelter, general support, financial assistance, were offered amply and unconditionally in the hours and days following the accident. However, great as this response was, this is not what is most striking about this accident. What is most striking about the accident is how lucky we were.

Who would ever think that, given the timing and the magnitude of the explosion, so many people, many fleeing with just the cloths they had on, would escape without serious injury? Few who have walked around that crater, seen the charred cars and the empty building foundations, would disagree with the conclusion that many there were saved only by a miracle. Unfortunately, miracles are a poor basis for public policy. You cannot count on them, and I am not about to count on them.

The fact is that there is no margin for error in today's pipeline industry. The natural gas industry does have an excellent safety record, especially when you consider that 25 percent of the energy that we consume moves through those pipelines. For example, there are seven major pipelines that cross our home State of New Jersey and hundreds of smaller ones. But the Edison accident never should have happened. We need to acknowledge Edison for what it is: a breakdown in the regulatory and safety program.

When the National Transportation and Safety Board testified before the Energy Committee in April at a hearing that I chaired, their analysis pointed nearly conclusively to multiple gouges on the pipeline as the probable cause of the disaster. These marks appeared to be due to some powerful machinery, such as a backhoe, that struck the pipeline repeatedly. At this point, we do not know whether the damage was unintentional or on purpose. We do not know who struck the pipeline or whether they might have been aware of the possibility. We do know, however, that there was no requirement of utility notification prior to the excavation, and we do know that there is no penalty for digging in the vicinity of the pipeline without notifying the utility operator.

Mr. Chairman, this is simply wrong, and this accident represents a failure of public policy. At the hearing before the Senate Energy Committee, every witness—I now emphasize "every witness"—agreed that we need a new national program of utility notification.

If someone is excavating or grading a site, there has to be a proper notification, and it has to be mandatory, not voluntary, without exceptions and with penalties for negligence or noncompliance. This national program will be created by comprehensive legislation that you are considering today.

Right now, the gas industry is making plans for rapid expansion into new markets, particularly in the areas of natural gas vehicles and electric power production. The Department of Energy has predicted that gas markets will expand by one-third over the next 15 years. If accidents occur, regardless of who is at fault or how the industry follows up, this growth will not occur. It is just that simple.

The telecommunications industry is likewise spending billions to expand its infrastructure and capabilities. If this investment, however, is held hostage by every backhoe operator in every State without serious controls and oversight, we simply will not see a lot of traffic on this information superhighway.

In one sense, this bill is unnecessary. Sooner or later, I predict, every State will adopt one-call provisions like those identified in the legislation. The reason is simple. Sooner or later, every State will experience a major accident involving third-party damage to underground utilities, like we did in New Jersey. Then, just as happened in New Jersey, one-call provisions will be introduced or strengthened.

This is not an issue of cost. Most States have these programs already. The problem is that absent sufficient political motivation these programs are just not as effective as they need to be. We should not have to wait for another disaster to understand the importance of this modest bill. The comprehensive one-call legislation represents a necessary step if we are to do everything reasonable and appropriate to protect the public from the kind of tragedy that struck in Edison.

Mr. Chairman, this bill obviously will not guarantee that another Edison will never occur. But mandatory, loophole-free, one-call programs based on a national model are a good place to start. Passage of this legislation will send a message to the public that our concern is serious and the risks a real. A national program will create a new level of awareness, and this awareness will be a powerful ally in our fight for increased safety.

Mr. Chairman, I encourage you in your work to pass this legislation. I believe there is no substantive reason why we cannot and should not act. I am ready to work with you and the committee to gain its approval and gain the approval of the whole Senate. We are prepared to make some modifications that have already been raised to it through discussions with the House of Representatives, and I hope we can work together to make this concept a public law in this Congress.

And I thank the chairman.

Senator EXON. Senator Bradley, thank you very much.

At this point, I am pleased to recognize my other colleague from the State of New Jersey. Senator Lautenberg and I have worked on many, many safety issues together. He is a key member of the Appropriations Committee on the whole matter of transportation and safety. Thank you for being here this morning, and we recognize you now, Senator Lautenberg.

STATEMENT OF HON. FRANK R. LAUTENBERG, U.S. SENATOR FROM NEW JERSEY

Senator LAUTENBERG. Thank you very much, Mr. Chairman. It is fitting that you and I are once again discussing an issue of safety. We have discussed it in terms of truck safety and other transportation matters, and this, as you are well aware, is part of the Transportation Committee's responsibility, as well as that of Senator Bradley, the Energy Committee. We are concerned about the delivery of gas throughout the pipelines throughout the country.

Mr. Chairman, thanks very much for accommodating our schedule this morning. I am chairman of the Superfund Subcommittee of Environment and Public Works, and they are about ready to start the hearing and I have to be there.

It is unfortunate that it took a disaster the magnitude that we saw in Edison, NJ, on March 23, to alert us that the Congress has to do something significant about this problem. You mentioned that Nebraska had an explosion and an incident. Once seeing those things, it reminds you of the fact that this is not unlike a wartime explosion. We were fortunate that there was such a small loss of life and the injuries were few. But it was, as Senator Bradley described it, something short of a miracle.

People tried to get out of the area, went to their cars, the cars were so hot from the flames that when they reached out to open the door their hands were burned. So, they were powerless. They had to run, just as Senator Bradley described, with the clothing on their backs and not much else, and it was quite a trauma for so many people to see what took place there and wonder at the fact that not only were so few lives lost, but the immediate leveling of home after home after home. This was a multiple dwelling community.

So, it would be unfortunate if, after having seen that experience, having had that experience, if Congress did not learn lessons and pass legislation to protect all Americans from this type of tragedy. Congress must not, and Senator Bradley and I will not, let the events of March 23 in Edison, NJ, fade into the past without taking corrective action.

As the committee considers one-call legislation it should keep in mind three points. One is mandatory participation, everybody has to participate, there can be no exceptions; swift and sure justice in the event of violation; and third, public education so people know that this process has to take place before any digging, any work around a pipeline, takes place. And the States which do have effective one-call laws, like Massachusetts, Minnesota, Arizona, require mandatory participation of all utilities, because they know otherwise that it is not going to work.

These laws cover all parties engaged in excavation, including the States' own public works agencies. They have mechanisms in place which apply swift and sure penalties to those who violate the one-call law, and they have effective public education campaigns. So, it works well.

S. 2101, the Bradley-Lautenberg one-call bill, was drafted with these three important points in mind—the House version of this bill, which has received extensive review, you mentioned it earlier, Mr. Chairman, also contains these three important elements: Strengthening one-call State requirements can prevent the devastation witnessed in Edison. The most common cause of pipeline damage is breaching of pipeline integrity by construction-related activity. That is sort of a refined way to say that when people work around there, whether they are utility companies, telephone companies, State public works agencies, if they touch that pipe, very often it is the weak spot that permits the gas to start to escape and then devastation follows.

So, Mr. Chairman, I thank you very much for holding this hearing, for your declaration that you are interested in moving safety legislation. We have seen that before. I strongly support the committee's efforts to address this issue. I offer my assistance in any way possible to make legislation mandating stronger one-call programs the law before the end of this Congress. And I thank you very much, Mr. Chairman.

[The prepared statement of Senator Lautenberg follows:]

PREPARED STATEMENT OF SENATOR LAUTENBERG

It is unfortunate that it took a disaster in Edison, New Jersey on March 23rd to alert many in the Congress to the significance of this problem. But, it would be even more unfortunate if Congress did not learn the lessons of Edison and pass legislation to protect all Americans from this type of tragedy.

Congress must not—and I will not—let the events of March 23 in Edison fade into the past without taking corrective action.

As the Committee considers one call legislation it should keep in mind three points:

- Mandatory Participation
- Swift and Sure Justice and
- Public Education.

Those states which do have effective one-call laws, like Massachusetts, Minnesota and Arizona, require mandatory participation of all utilities. These laws cover all parties engaged in excavation, including the states' own public works agencies. They have mechanisms in place which apply swift and sure justice to those who violate the one-call law and they have effective public education campaigns.

S. 2101, the Bradley/Lautenberg one-call bill, was drafted with these three important points in mind. The House version of this bill, which has received extensive review, also contains these three important points.

Strengthening one-call state requirements can prevent the devastation witnessed in Edison. The most common cause of pipeline damage is breaching of pipeline integrity by construction-related activity.

I strongly support the Committee's efforts to address this issue and offer my assistance in any way possible to make legislation mandating stronger one call programs law before the end of this Congress.

Thank you.

Senator EXON. To both of you, thank you very much. I know your schedules are very tight and I will not hold you up. I just want to ask a question which either one or both of you may answer. I am sure this is going to come up somewhere along the line. First, I would like to correct any misimpression I may have created that the situation in Nebraska involved pipelines. The incident concerned cable and telephone wires. It points to the extent of the problem, though, which we have. With the new information highway system, I imagine we will have many more wires spread around the United States in one form or another, and there will be much more digging. We must have a better system than we have had in the past.

I would hope that sometime today someone in panel 1 or 2 could answer a question I have: with all of the sensing devices we have today that pick up a penny in the ground someplace, why can't we or shouldn't we develop some kind of a device which would see what is down there before digging? It could be required to be used in addition to one-call, of at lease commercial operators.

My one question is, with all the concern today in Congress, among the Governors, and States with regard to mandates from Washington, DC, in your bill, are there mandates that would require costs at the State or local level? If not, fine. If so, how do we bridge the problem of unfunded mandates? To some extent, each and every one of us in the Senate have taken the pledge of no more mandates without paying for them. Is this a problem in your bill, or not?

Senator BRADLEY. Mr. Chairman, it is not. The bill does not create new Federal or State-run programs. Many one-call centers are currently run privately, and H.R. 4394, which is the House version that we have talked about modifying our bill to mirror, would not upset the situation. States are not obligated to establish and run a program.

States would be required to pass laws that require all underground facility operators to participate and all excavators to call before digging and establish a workable penalty and enforcement system. But it does not provide any financial sanctions such as a loss of highway funds. That is what we had in our bill initially. That is now gone.

It does make States responsible for enforcement actions. Many States have addressed this issue of enforcement already. It does require municipal governments to mark their known water and sewer lines. It does provide incentives: $4 million a year for the next 3 years in grant money to assist States and existing one-call centers to improve their programs. And it does emphasize excavator training and public awareness.

So, the answer would be "No" on mandates.

Senator EXON. Well, let me just follow up very briefly, Senator, then, certainly there is going to be some expense involved in this. Are you suggesting that the utilities bear the cost, essentially most of the cost, for establishing a one-call system? Somebody is going to have to be on the line to answer the phone. Who pays for that?

Senator BRADLEY. Well, this cost is a minimal cost, to say the least, and we have $4 million in the bill that would provide incentives for existing one-call centers to improve their programs. So, there would be Federal assistance.

Senator EXON. Would not the utilities be involved?

Senator BRADLEY. They could be. They could be. But at the moment we think it is in everybody's interest to have participation from all sources here.

Senator LAUTENBERG. Mr. Chairman, if I might add a note, I have another piece of legislation that we are working on now that makes requirements of companies, as well as Government, to participate more fully in inspections. You mentioned that in your comments. There is an ability to run something called a pig through these pipes and it detects very accurately the smallest breach of structure.

We also are asking the Federal Government, the Department of Transportation, to have a mapping program that lays out all of the pipelines in the country. Unfortunately, there are many places where we do not know what is going on. When we combine that with the one-call bill that Senator Bradley has put together with me, we would be able to deal with this, I think, in a manner in which the costs would be distributed in such a way that they are really de minimus to the States.

There is a $12 million 3-year program that would provide incentive grants for the States to participate in this program. And I do not think that this comes under the classification of a mandate that would lay any other costs upon the States that they would not be taking to protect their citizens in the first place.

Senator BRADLEY. Mr. Chairman, if I could add, you mentioned that some of the most effective one-call programs in the country are now essentially utility run, or utility financed programs. There could be some increased utility contribution.

As to the hammer, there no longer is the loss of highway funds. We basically allow the Secretary, or a person injured by a third-party accident, to file civil suits against the State in order to compel a better program. That is the only sanction that we have.

Senator EXON. Senator Bradley, Senator Lautenberg, thank you very much, and you are excused.

Senator BRADLEY. Mr. Chairman, thank you.

Senator EXON. I would ask at this time that the entire prepared statements by Senators Bradley and Lautenberg be included in the record at the appropriate point.

Also, at this point in the hearing, I would like to ask unanimous consent that a statement by Congressman Pallone be entered into the record.

Also, I ask that all of the written statements by the witnesses appearing on panel 1 and panel 2 be included in the record at an appropriate point.

I have just been advised that Congressman Pallone is in the room. Congressman, please come forward at this time. I did not know you were here. I have included your statement, assuming you could not make it. I understand that your colleague, Congressman Sharp, will not be with us, and if he has a written statement, I also ask unanimous consent that it be included in the record.

[The prepared statement of Congressman Sharp follows:]

PREPARED STATEMENT OF MR. SHARP

Thank you Mr. Chairman, for the opportunity to appear before you today to discuss S. 2101, the "Comprehensive One-Call Notification Act of 1994," and the substitute amendment to its House companion, H.R. 4394, that I will offer at my Subcommittee markup tomorrow.

As you know, this bill, introduced by Senator Bradley and by Mr. Pallone in the House, came about in the wake of the tragic accident that occurred in Edison, New Jersey on March 23, 1994. Although investigations are ongoing, it appears as though the accident was a direct result of excavation damage. No one can say for certain whether or not an effective one-call system would have prevented the damage that occurred on that Texas Eastern Pipeline, however, it would have certainly reduced the risk. It is clearly evident that we need to pay attention to the causes of such incidents, and do what we can do curb this type of activity in the future.

Effective one-call notification systems that meet minimum standards throughout the country, will take us a step closer to that goal.

One-Call systems are designed to help prevent the occurrence of damage to underground facilities such as pipelines, fiber optic cables, and electric lines and to prevent injury or death to excavators performing work in the vicinity of such facilities. Responsibilities fall on the excavator, who must make the call and wait the appropriate time before digging, and the underground facility operator, who must properly locate and mark their underground facilities in a timely manner. The one-call center serves as a clearinghouse for this exchange of information.

While most states currently have some sort of one-call system in operation, it is true that many of these systems are inadequate and inconsistent throughout the country. H.R. 4394, and my substitute amendment, attempt to address this by establishing uniform, minimum standards that States must follow. Conversely, it is also true that many States have very good programs on which they have spent a lot of time and money to develop and promote. My amendment therefore gives States the flexibility they need to continue running their effective programs as they currently exist. For States with programs that are at least as protective of public health and safety as the program that is outlined in this legislation, there is a provision that allows for self-certification with the Department. It is then up to the Secretary to disapprove of such a program.

This legislation does a lot to encourage and promote the development of one-call systems, particularly those that go beyond minimum standards. The Secretary has the authority to provide incentives in the form of grant money ($12 million over three years) for the establishment and improvement of one-call systems. The development of a nationwide toll-free phone number will enable public education through the use of public service announcements and other literature. A yearly workshop to be help by the Department of Transportation in conjunction with State and focal governments, underground facility operators, excavators, and one-call system operators, will ensure the development of a model program and will raise awareness and involvement of all of these entities as we strive to improve these systems throughout the country.

It is essential that we work to move this legislation in the remaining days of the session. This bill enjoys widespread support from industry, government, utilities, and one-call system representatives. The time is right to adopt nationwide minimum standards for one-call. The near-tragedy that occurred in Edison, New Jersey last March should serve as a national, not just a local wake up call. We don't want to have to wait for 50 accidents in 50 states before we all benefit from the lessons learned from this accident.

Senator EXON. I apologize for not being advised you were here. Thanks for coming over. I recognize you for any statement you have at this time.

STATEMENT OF HON. FRANK PALLONE, JR., U.S. REPRESENTATIVE FROM NEW JERSEY

Mr. PALLONE. Thank you, Mr. Chairman. I will try not to repeat what the two Senators from my home State already said to you.

I just wanted to thank you for giving me the opportunity to testify this morning, and also having the hearing on Senator Bradley's one-call notification bill, and as the Senator mentioned, I have introduced companion legislation in the House of Representatives and have been active on the issue of pipeline safety since the accident took place in my district in Edison, NJ, and obviously your attention to this problem is greatly appreciated.

I listened to what you said earlier, and I would like also to thank Congressman Sharp, who has been working hard as the subcommittee chairman to move this legislation in the House. The concept of the nationwide one-call notification system has basically been discussed for about 20 years, but it did not come to the forefront of my legislative agenda until this major pipeline exploded in Edison.

I think that the two Senators basically described the incident and the damage that occurred from it, but I just wanted you to know that it was so powerful that nine apartment buildings were leveled and the fireball shot into the sky was seen in three States, it was that powerful.

I visited the scene the next day and was really shocked by the total devastation there. It really looked as though a bomb had blasted the buildings to smithereens, though fortunately nearly all the residents of the apartment buildings survived. We only had one fatality, which was a woman who died of a heart attack as she was fleeing the scene. It was really a miracle that we did not have more.

We have been working since the accident, as you know, both Senators, Congressman Sharp and myself, to prevent this kind of disaster from recurring, and one lesson we learned is the fact that this explosion, like the majority of the pipeline explosions, was likely due to third-party damage, so the bill you have before you today as well as the House version would basically work toward prevention of the third-party damage by requiring all States provide for the mandatory compulsory one-call notification provision.

Now, the concept is simple. Before digging, the excavator calls the central number to learn if there are any underground facilities. If there are, the facility operator comes out to the site to mark the ground, and the excavator avoids hitting the line because its location is known.

One-call systems can work, but I believe that mandatory participation, effective enforcement, and public education are essential, and these are the three prongs, if you will, of this legislation that we have included.

Many of the State statutes that are currently in force lack these three important elements. The legislation allows for immediate citations for violations. The citations, which can be written on the spot like a traffic ticket, are a swift means of enforcement that will show violators that failing to call or mark is serious and costly.

We have a public education provision, because I believe that without those, one-call programs will not reach their full potential, and we also provide advertising, the telephone number in order to reach small-scale excavators as well as the large, established companies.

Now, I know you mentioned the costs, and I would say that the costs associated with one-call systems are minimal. Most of the time, the underground facility operators will fund the program, and sometimes with the help of fees paid by excavators. Facility operators usually find one-call systems to be of great benefit in avoiding accidents, and feel that paying the fee generally is worthwhile in order to prevent damage to their lines. The charges are usually the result of negotiation among facility operators.

If I could say, Mr. Chairman, I wanted to note that in the legislation, or at least the House version that has been revised somewhat now, we actually leave flexibility up to the State so that if States do not want State facility operators, or local facility operators, meaning municipal government or government entities that are facility operators for some reason to have to pay charges, they do not have to, so theoretically, there would not have to be any cost to any government unit, if that is the way the State decides to go.

I just wanted to say that this issue attracted a lot of national press coverage at the time when the explosion took place. I really think it is crucial now in the aftermath, in that there perhaps is not as much press interest in it, but the bottom line is, it is very crucial to pass this legislation, because I think that without it we could very easily see accidents like this occur again, and in fact since March there have been accidents in Wyoming, even in Washington, DC, where a line was hit at the National Airport, so this is not something that is isolated. It is something that is not going to go away, and this is a major way to prevent, I think, a lot of these accidents from occurring.

Thank you very much.

[The prepared statement of Representative Pallone follows:]

PREPARED STATEMENT OF MR. PALLONE

Mr. Chairman, I would like to thank you for giving me the opportunity to testify this morning and for holding this hearing on Senator Bradley's One Call Notification bill. As you know, I have introduced companion legislation in the House of Representatives and have been active on the issue of pipeline safety since the accident in Edison, New Jersey. Your support and attention to this problem is greatly appreciated. I would also like to thank both Senator Bradley and Senator Lautenberg for their work on this issue and Congressman Sharp who is working hard to move this legislation in the House.

The concept of a nation-wide one-call notification system has been discussed for more than 20 years. It did not come to the forefront of my legislative agenda, however, until the a major pipeline exploded in Edison, New Jersey on March 23, 1994. The damage caused by this blast was so powerful that nine apartment buildings were leveled and a fire ball shot into the sky that was seen in three states. I visited the scene the next day and was shocked by the total devastation there—it looked as though a bomb had blasted the buildings. Fortunately, nearly all of the residents of Durham Woods apartment buildings survived the accident without severe physical harm, (one young woman died of a heart attack), but hundreds were left homeless.

During the months since the accident, as the Durham Woods residents who were left homeless have been working to rebuild their lives, the Senators, Congressman Sharp, and I have been working to prevent this kind of disaster from happening in other communities. One lesson we learned from this accident is the fact that this explosion, like the majority of pipeline explosions, was likely due to third party damage. S. 2101 and the House version of this bill, would work toward prevention of third party damage by requiring that all states provide for a mandatory, compulsory one call notification system.

The concept is simple: Before digging, an excavator calls a central number to learn whether there are any underground facilities at the excavation site. If there are, the facility operator comes out to the site to mark the ground. The excavator avoids hitting the line because its location is known. This common sense approach prevents unnecessary deaths and damage to utilities which are a direct result of simple ignorance. One call systems provide information that could not otherwise be determined. An excavator cannot know what is buried beneath his feet, and a facility operator cannot know who is digging miles away from his office unless this information is provided to him. One call systems are the vehicle for passing this crucial safety information along.

One call systems can work this way, but mandatory participation, effective enforcement and public education are essential. Many of the states statutes that are currently in force lack these important elements. Without mandatory participation, excavators have no insurance that they are digging in a safe place, and facility operators are not assured that their lines will not be hit. If all excavators and facility operators participate, without exemption, these simple phone calls and routine markings will prevent property damage, death, injury, and the need for expensive repairs.

Because many states have complex enforcement mechanisms for their one call laws, violations often go unpunished. This legislation will allow for immediate citations for violations. These citations, which can be written on-the-spot, like traffic tickets, are a swift means of enforcement that will show violators that failing to call or mark is serious and costly.

Of course, without public education, one call programs will not reach their full potential. We must advertise the telephone number and reach the small scale excavators as well as large established companies. As the public becomes acquainted with these systems, and calling becomes routine for excavators, I feel confident that we will see fewer and fewer of these unnecessary tragedies.

The costs associated with one call systems are minimal. The underground facility operators generally fund the program, sometimes with the help of fees paid by excavators. Facility operators usually find one call systems to be a great benefit in avoiding accidents, and feel that paying the fees is worthwhile in order to prevent damage to their lines. The charges are usually the result of negotiation among facility operators, excavators and one call centers, and often are scaled to reflect the facility's presence in the state.

The national press coverage of the devastating scene attracted public attention to this issue. We must now take the explosion that awoke the residents of Durham Woods in the middle of the night as a wake up call to us to deal with the issue of pipeline safety. I am confident that mandatory, compulsory one call systems will work prevent accidents and tragedies like the one in Edison.

Finally, the case the accident in Edison was not an isolated incident. Since March we have seen accidents in Wyoming (where two state highway workers were killed), and, on the day after Mr. Sharp's hearing on this issue, in Washington, D.C. (where a line was hit at National Airport). Let us act now, we do not need to wait for every state to experience a tragedy like Edison. Thank you, Mr. Chairman.

Senator EXON. Thank you very much, Congressman. I have no questions. I guess you heard my question about the old situation of mandating to lower levels of government. Do you endorse the general concept, in answer to my question, as was stated by the two Senators on that issue?

Mr. PALLONE. Yes, and as I said, there does not have to be any cost to local governments. I mean, basically the cost could be incurred by the facility operators, and they could even exempt themselves in cases where the government was the facility operator. I am not suggesting that is the way to go, but that is allowed pursuant to the legislation.

Senator EXON. Congressman, thank you very much for being here. We appreciate your coming over and your valuable input on this issue. We are going to be working together to try and resolve this very legitimate safety issue.

When I hear you and the two Senators talk about the total devastation of this blast, and the fact that only one fatality occurred, it just must be one of the major miracles in recent times. We are fortunate, indeed, that this terrible accident did not take more lives. Were there any serious injuries in connection with this?

Mr. PALLONE. There were some injuries, but my understanding, and I am not sure anyone really knows for sure, but apparently the shock from the explosion, the noise from the explosion woke the people up in the complex, and they had about 7 minutes before the fireball actually reach the apartments, and there was a berm next to the asphalt plant where this occurred, so apparently in that 7 minutes people were able to flee, and it was amazing, Mr. Chairman, because they helped each other. In other words, rather than just run out of the houses, they went around and knocked on each other's doors, and there were some people that were disabled that had to be helped out, and amazingly they got away in that 7 minutes or so. Once they got about 1,000 feet or so away, they were OK. I mean, there were some injuries, but other than the one woman who died fleeing the scene from a heart attack, that was pretty much it.

Senator EXON. Just for my information, how far was it from the point of the blast to the apartment building?

Mr. PALLONE. I am not very good at feet, but I can tell you it is probably the distance from here to maybe the Hart Building. That is all. It is not that far, I do not think. I mean, just from looking at it.

Senator EXON. It is hard to imagine.

Mr. PALLONE. There was a berm in the way, a very large berm that deflected a lot of it, maybe a little further than that, but it was not that far. I mean, I do not think it was, say, as far as the Capitol from here, looking at the area.

Senator EXON. Thank you very much, Congressman. You are excused.

At this time, let me call panel No. 1: Jim Hall, Acting Chairman of the National Transportation Board, and Cesar DeLeon, Deputy Associate Administrator for Pipeline Safety, Research and Special Programs Administration, Department of Transportation. Please take your places. I hope you have heard that I have already included in the record the full text of your statements. We would appreciate your summarizing these statements at this time. Thanking you for being here, we will start with Mr. Hall.

STATEMENT OF JIM HALL, ACTING CHAIRMAN, NATIONAL TRANSPORTATION SAFETY BOARD; ACCOMPANIED BY CHARLES BATTEN, PIPELINE DIVISION

Mr. HALL. Good morning, Mr. Chairman. Before I begin my remarks, I would just like to inform the committee that at 3:45 a.m. this morning, Amtrak train 49, en route from New York to Chicago, derailed near Batavia, NY. There were 140 passengers and crew on the train, according to initial reports.

Board member John Lauber is heading a team of 10 investigators from the Safety Board that is proceeding now to the scene of that accident, and it will be the sixth Amtrak accident that we have under investigation in the last 12 months.

Senator EXON. I had not heard of that before. Was this a serious accident, with fatalities, or do we know at this time?

Mr. HALL. At this time, sir, we do not know of fatalities. It is a serious accident. I think the majority of the train cars, according to initial reports, were derailed.

Senator EXON. This happened where, Mr. Hall?

Mr. HALL. In Batavia, NY, sir.

Senator EXON. And that is—I am not familiar with the geography—is that north or south of New York City?

Mr. HALL. It is western New York State.

Senator EXON. Thank you very much. Please proceed. I appreciate the information.

Mr. HALL. It is a pleasure for me, Mr. Chairman, to appear on behalf of the National Transportation Safety Board. Accompanying me is Charles Batten of our Pipeline Division, who served as our investigator in charge of the Edison, NJ, accident.

I am sure it comes as no surprise that the Safety Board is appearing today to express support of the Comprehensive One-Call

81-903 O - 95 - 2

Notification Act, known as Senate bill 2101. Almost since the beginning of the Safety Board in 1967, we have pushed for ways to reduce excavation-caused pipeline accidents. In fact, in 1972, the Safety Board conducted a symposium on excavation damage, a symposium which led to the establishment of today's one-call notification network.

When we held the 1972 symposium, there were two or three one-call systems in the country, and each covered a city. Today, the situation is quite different. Now, all but one State have one-call notification systems and utility location and coordination councils operate in most communities, so substantial progress has been made, yet as the Senators and Representatives testified earlier, the potential of this one-call excavation notification system has not been realized, because many buried facility owners are exempt from coverage, and some excavators fail to participate in the voluntary system.

So, legislation is needed, and the pending measure will, we believe, strengthen the one-call system by calling for the full—and I emphasize the word "full"—participation by all organizations, whether public or private, that operate buried facilities or that excavate.

Provisions specifying conditions for the use of one-call notification centers, and endorsing the use of administrative penalties against system violators, along with other measures, will significantly broaden the effectiveness of excavation damage prevention programs.

My submitted testimony, Mr. Chairman, cites several ways the legislation can be enhanced, and I will not repeat those remarks, but I do want to mention our belief that too much time is afforded for State compliance, because States already have ongoing excavation damage prevention programs which can be readily adapted to meet the legislation's requirement.

Simply put, the Safety Board feels the legislation will go a long way in reducing damage to pipelines and other buried facilities. At the same time, we need to bring attention to the overall problem of excavation damage, and this need has prompted the Safety Board to again conduct a 2-day workshop to identify new ideas and approaches for preventing excavation-caused damages. The workshop will be held September 8 and 9 here in Washington, and initial indications are that it will be well attended by all involved with the issue. We are also pleased that the Office of Pipeline Safety is cohosting the workshop.

Action on the problem of buried facility damages can only benefit the public. While pipeline damage often receives the bulk of the attention, other forms of excavation damage also affect safety and economy. Excavation damage, as you mentioned earlier, to telecommunication cables, can pose severe safety and economic consequences if, for example, emergency 911 services and economic data transmittals are disabled.

Just last week, MCI long-distance service between New York and Washington was disrupted for nearly 6 hours after construction workers in West Windsor, NJ, sliced through a lead-encased transmission cable. Damage to water and sewer underground systems also impact both citizens and the environment.

These are just a few examples which we believe add evidence justifying the need for one-call legislation and other excavation damage prevention programs.

Mr. Chairman, that concludes my remarks, and Mr. Batten and I would be glad to respond to any questions the panel may have.

[The prepared statement of Mr. Hall follows:]

PREPARED STATEMENT OF JIM HALL

Good morning Mr. Chairman and Members of the Subcommittee. I appreciate the opportunity to appear on behalf of the National Transportation Safety Board to discuss S. 2101 and H.R. 4394, legislation aimed at preventing excavation damage to our Nation's infrastructure, especially buried pipelines.

EXCAVATION/THIRD PARTY DAMAGE

Excavation-caused damage to pipelines is the largest single cause of pipeline accidents today. Of the accidents reported to the Office of Pipeline Safety in the Research and Special Programs Administration (RSPA) in 1993, at least 26 percent of the 119 distribution system accidents, 16 percent of the 97 gas gathering and transmission accidents, and 22 of the 229 liquid pipeline accidents were caused by excavation errors. Of the accidents reported in 1992, 34 percent of the 103 gas distribution system accidents, 14 percent of the 75 gas gathering and transmission accidents, and 18 percent of the 223 liquid pipeline accidents were caused by third party damage. Our review of Office of Pipeline Safety reports found that excavation caused damages amounted to more than $17 million in 1993 and $31 million in 1992. However, other reports either contained insufficient information for use to determine if excavation damage to pipelines was involved or to determine the value of the property damage losses even though the accident was caused by excavation.

The effects of outside force damage pipeline accidents are most severe in the gas distribution systems. Those accidents account for most of the injuries and fatalities in gas distribution accidents and most of the environmental damages caused by product releases from liquid pipelines.

The Safety Board is consequently very concerned about this issue and welcomes this hearing. This morning we will discuss several accidents that highlight existing problems and demonstrate the need for new safeguards to reduce excavation, or so-called third party damage to pipelines.

On March 28, 1993, at Reston, Virginia, more than 400,000 gallons of diesel fuel was released into Sugarland Run Creek and the Potomac River, when a 36-inch-diameter petroleum products pipeline ruptured. Although no one was injured, there was substantial environmental damage. The Safety Board's investigation revealed that the pipe ruptured because of metal fatigue in a crack initiated by mechanical damage from heavy construction equipment at an undetermined time. Despite a thorough search of construction records for the seven years prior to the accident, we were unable to determine when the damage may have occurred.

Our investigation found that the pipeline company involved was an active participant in the area's one-call notification system. Moreover, our investigators found that the State of Virginia required excavators to give 48 hours notice in advance of any planned excavations.

Excavation damage also arose in an accident in June 1993, in Cliffwood Beach, New Jersey, where three residents died and three other individuals were injured when a house exploded. In this accident, a contractor working for the local gas distribution company informed the one-call notice center of the date when excavation was to begin. Excavation notification in New Jersey at the time of the accident was voluntary. Before beginning work, the contractor was provided information on the gas line's location, but not on the gas service line locations in the area. The contractor was installing new gas lines when his equipment struck and severed an unmarked, inactive steel gas service line. The contractor installed a temporary repair on the visible break but he did not check for other possible damage. The Safety Board found a second break in a ¾-inch inactive service line at a fitting within a buried, paved-over foundation of a previously demolished house occurred. When gas pressure to the service line was restored, gas then migrated into an occupied house and exploded.

Another fatal accident occurred on July 22, 1993 in St. Paul, Minnesota, killing three people and injuring 12 others. A city public works employee severed a 1¼-inch plastic service line with a backhoe at a joint adjacent to a building foundation wall. Gas entered and ignited in the building. The Minnesota one-call notification center had been notified of the construction and gas company personnel had accurately marked the location of the pipeline before construction began. Minnesota State law required notice before excavating, and it applied to all excavators and all operators of buried facilities.

Three other accidents which have occurred this year further graphically demonstrate the ramifications of third-party pipeline damage.

On March 23, 1994, a 36-inch-diameter steel high pressure gas transmission pipeline in Edison, New Jersey, ruptured. The escaping gas ignited, sending flames hundreds of feet into the air. Heat radiated from the natural gas flames and ignited several apartment buildings. Eight apartment buildings were destroyed by fire, and more than 1,500 residents of the apartment complex were evacuated. Dozens of residents and some emergency response personnel were injured, treated at a local hospital and released.

On inspection of the pipeline, our investigators saw that the pipe had been damaged by at least two different types of mechanized excavation machines. The time of the damage is yet undetermined. Our investigators have learned that the transmission company had not been notified of any excavation activity in the area, even though the company was an active participant in New Jersey's voluntary one-call notification system.

In another accident investigated by the Safety Board, a 10-inch-diameter natural gas gathering pipeline near Green River, Wyoming, was ruptured on May 3, 1994, when excavation equipment operated by contractor employees struck and damaged the pipeline. The two employees operating the equipment were killed and two others injured, one of whom died from burn injury complications more than a month after the accident. There was a one-call notification center operating in the area. However, the Wyoming damage prevention law did not require participation in the notification system and the pipeline operator did not participate in the notification system.

The most recent 1994 accident occurred June 9 in Allentown, Pennsylvania. In this accident, a 2-inch-diameter natural gas service line was pulled apart at a coupling less than four feet from an eight-story senior citizen apartment building. The coupling was separated on the side nearest the building by a distance of about 2.5 inches. Another coupling on the line exhibited signs of being pulled about a quarter inch, but it was not completely separated. There were numerous gouges and tears in the pipe's coating on a portion of a 59-foot section of pipe, and the pipe was permanently deformed and removed for further examination. One person died in this accident, one was critically injured, and almost 100 people were injured.

Adjacent to a portion of the pipeline, an old leaking fuel tank, its concrete pad and contaminated soil were removed by a contractor. The excavation was reportedly being backfilled. The pipeline was exposed in the excavation area, and testimony and pictures conflict as to whether the excavated line segment was being supported. The bucket for the track hoe was observed in the excavation, and the Safety Board is investigating to determine whether the pipe was pulled out of the coupling due to the pipeline being unsupported, being struck by the track hoe bucket, or as a result of the trench sidewall caving into the excavation.

Gas migrated from the coupling to the building's boiler room. A large vacant area existed under the boiler room floor to accumulate fresh air to the boilers. Grates were in the side outside walls and through the boiler room floor. There were vertical shafts from the boiler room to all floors in the building for electrical conduit and a trash chute. This allowed gas to go up through the building. Safety Board investigators found damage on all floors as a result of gas migrating through the building which ignited and exploded. Some of the most severe damage was on the fifth through seventh floors, and reinforced concrete on the bottom of the first floor was moved upward almost a foot in some locations by the explosion's force.

The type of accidents described above, unfortunately, are not new occurrences. Almost since the beginning of the Safety Board in 1967, we have investigated excavation-caused pipeline accidents, and based on these investigations, recommendations have been issued calling for remedial action.

The accident investigations in the late 1960s and early 1970s prompted the Safety Board to hold a symposium in 1972 on preventing excavation-caused damages. That symposium brought together representatives of organizations involved with excavation and pipeline operations to explore solutions to the problem and to improve public safety. As a result of the symposium and the subsequent recommendations issued by the Safety Board, the American Public Works Association today serves as the primary support for local, state, and national groups dedicated to preventing excavation damage to our Nation's pipeline infrastructure.

When we conducted our 1972 symposium, only two one-call excavation notification systems—one in Portland, Oregon, and the other in Rochester, New York—existed.

All but three States currently have one-call notification systems, with most States providing statewide coverage through a single one-call center. Today, Utility Location and Coordination Councils composed of operators of underground facilities, excavators, and city, county, and state agencies now operate in most communities. These councils coordinate planned major construction projects to minimize pipeline damages. In 1972, only a few of these councils existed. A national council of ULCCs now exists to provide general guidance to many local and state organizations. There were no State laws on excavation damage prevention in 1972, but today laws exist in all but seven States.

Clearly progress has been made. The voluntary, cooperative actions of buried facility operators, state lawmakers, and excavators have been most responsible for reversing the increasing excavation-caused accident trend experienced in the 1970s. As a result of these safety advances, many lives have been saved, injuries prevented, and countless dollars in property damage avoided. It is relevant to note that these improvements came at a time when there was substantial growth in residential, commercial and highway construction, and extensive urban renewal projects in most barge cities.

Once most of the States implemented one-call notification systems, activated underground location and coordination councils, and implemented damage prevention laws, the frequency of excavation-caused pipeline accidents markedly reduced. Unfortunately, that rapid reduction trend began to fade in the 1980s, and the downward trend has stagnated. The pipeline accident statistics since 1985 show that the average number of outside-force-caused pipeline accidents has remained relatively constant. But further decreases are needed as the discussed Safety Board investigations demonstrate.

Safety Board staff consequently initiated talks with the various accident prevention organizations to explore ways to enhance existing programs. In these discussions, staff learned that many of the original damage prevention objectives have yet to be achieved. For example, the full potential of one-call excavation notification systems have not been realized. This is because many buried facility operators, primarily government-owned water, traffic control, and sewer systems, do not participate in active on-call centers. Consequently, contractors and excavators who want to learn of infrastructure locations to minimize damage must not only call the one-call center, but also must individually call all non-participating operators. Furthermore, there are agencies and organizations performing excavations that do not bother with notifying anyone of their planned excavations. Safety Board staff learned that State agencies, railroads, and Federal county, and local government units often obtain statutory exemptions from notification requirements; yet, their actions pose the came threats to the infrastructure as do other excavators. Additionally, the penalty provisions in most state laws are not invoked because they are cumbersome and costly to administer. As such, nationally there have been few instances where a person has been penalized for violating any provision of state damage prevention laws.

The Safety Board consequently believes these programs must be strengthened. Although our discussions are not finished, we believe we have identified several key components for successful damage prevention programs. These program elements require:

- Full participation by all organizations that operate buried facilities, whether public or private;
- A true, statewide one-call notification center with specific instructions specifying the conditions for its use; and
- Monetary administrative penalties against violators, injunctive relief against habitual violators and imminently dangerous excavation operations, and criminal action against persons whose actions violate the law and result in casualties or substantial property damage.

In addition, the Safety Board, in cooperation with the Department of Transportation's Office of Pipeline Safety, will hold a two-day Workshop beginning September 8, 1994, to identify additional ideas and approaches for preventing excavation-caused damages to buried facilities. The Workshop will provide experts the opportunity to come together to develop new excavation safety initiatives and to identify current accident prevention activities that should be expanded nationwide.

We will keep you advised of developments regarding the Workshop, and a copy of the proceedings will be forwarded to you when it is available.

Mr. Chairman, the Safety Board favors passage of one-call legislation, and we offer for your consideration several ways the legislation could be further enhanced to afford additional protection for pipelines and the public.

Most States already have a statewide one-call notification center and have excavation damage prevention statues that could be amended readily to bring state programs into compliance with the provisions in the legislation. We therefore believe that too much time is afforded for State implementation.

The legislation should clarify that the States are to establish excavation damage prevention programs, not one-call notification systems. As mentioned earlier, all existing one-call notification systems have been established through voluntary, cooperative relationships among operators of buried facilities. Their structures and operating conditions have been developed to address local conditions unique to the organizing agencies. The Safety Board believes that flexibility has been one of the more important factors in the successes achieved. We therefore encourage the Committee to retain as much flexibility as possible relative to the organization structure and the day-to-day operations of one-call notification centers.

To avoid confusion and duplication of effort, the provisions applicable to one-call notification center operations and those applicable to state governments should be clearly distinguished. This could be accomplished by using the term "excavation damage prevention program" when referring to actions to be taken by a state and using the term "one-call notification center" when referring to the system used to receive and disseminate information on planned excavations.

There should also be a requirement for providing educational opportunities about the damage prevention program for excavators and buried facility operators. Information on the state damage prevention statute and use of one-call notification system could help broaden its use. In this vein, training and licensing provisions for operators of mechanized excavation equipment should also be considered by the states.

These are just a few of the improvements the Safety Board believes warrant attention in any pipeline safety protection initiative. Our accident investigations offer compelling testimony for further pipeline safety enhancements and we are willing to work with the panel in any way possible.

That concludes the Safety Board's prepared remarks and I would be pleased to answer any questions you may have.

Senator EXON. Mr. Hall, thank you very much. We appreciate your concise statement and the support you are offering. Now, let us call on Mr. DeLeon. We are delighted to have you here this morning, and please proceed. Your full statement has been included in the record.

STATEMENT OF CESAR DELEON, DEPUTY ASSOCIATE ADMINISTRATOR FOR PIPELINE SAFETY, RESEARCH, AND SPECIAL PROGRAM ADMINISTRATION, DEPARTMENT OF TRANSPORTATION

Mr. DELEON. Thank you. Good morning, Mr. Chairman.

I am Cesar DeLeon, Deputy Associate Administrator for Pipeline Safety in the Research and Special Programs Administration. Since I am new to this position, this is my first opportunity to testify before your committee. Thank you for the opportunity to appear before you today to discuss issues concerning protection of our pipeline infrastructure—one of the most important, yet least understood transportation systems in the Nation. In protecting the integrity of pipelines, we protect the health and welfare of the people and the environment in proximity to those pipelines. The issues we will discuss today have been framed in legislation that has been proposed by your colleagues from New Jersey, Senator Bradley and Senator Lautenberg, and in the House by Congressmen Pallone and Sharp.

Most of my statement will focus on this legislative initiative, but I believe it is important that I put this initiative and the issues it confronts in the context of the entire pipeline safety program.

Mr. Chairman, with a growing list of mandates for action from Congress and the National Transportation Safety Board, and with resources incapable of keeping up with those mandates, the pipeline safety program has had to be one of crisis management rather than risk management.

However, Secretary Peña has committed the Department to pursue a budget for pipeline safety in fiscal year 1995 that is double what it was in fiscal year 1993, signaling a new era for pipeline safety. This will be a new era in which the pipeline safety program can become leader in risk management and be a credible, technically solid and independent assessor of the pipeline infrastructure and the actions of industry to protect and operate that infrastructure. I believe that the proposed legislation can be an effective tool to help us realize that purpose.

Mr. Chairman, you have asked for our views on proposals contained in S. 2101, the Comprehensive One-Call Notification Act of 1994. Our comments are based on the July 14 draft companion bill, H.R. 4394. The legislation addresses the single leading cause of pipeline failures, external force, or third-party damage which is typically caused by heavy construction equipment striking a pipeline. It is predicated on several concepts that we fully support, including a comprehensive statewide one-call notification system, and a Federal pipeline safety grant program that provides funds to the States to encourage State adoption of effective one-call programs.

The essential elements of an effective one-call system are the mandatory participation by all underground facility operators, only narrowly drawn exceptions to a general requirement that all excavators call before they dig, effective, easily administered sanctions, 24-hour coverage with provisions for emergency excavation, development by the Department of Transportation of a model one-call program, development by the Department through partnerships of a nationwide public education program, an enforcement provision that leaves discretion to the States in establishing penalties and provides for increased penalties when damage results from a violation, and last, Federal implementation of a nationwide, toll-free one-call telephone number.

We believe that several concepts in the bill, including the nationwide telephone number, the grant provisions and the model program, need clarification to assure that they accomplish their objectives. Details of these concerns are included in my written statement. We will continue to work with the committee staffs to resolve these items.

Mr. Chairman, we believe that Congress has made great progress toward crafting one-call legislation that will significantly enhance our ability to protect the integrity of pipelines. We will continue working closely with the committees to produce an effective bill.

Thank you. I would be happy to answer any questions, Mr. Chairman. Thank you.

[The prepared statement of Mr. DeLeon follows:]

Prepared Statement of Cesar DeLeon

I am Cesar De Leon, Deputy Associate Administrator for Pipeline Safety in the Research and Special Programs Administration (RSPA). Thank you for the opportunity to appear before you today to testify on behalf of Secretary Peña, Acting Administrator Ana Sol Gutiérrez, and the Department of Transportation (DOT) concerning key elements of a risk management regime to protect the Nation's pipeline infrastructure. In protecting the integrity of pipelines, we protect the health and welfare of the people and the environment in proximity to those pipelines.

The issues we will discuss today have been framed in legislation that has been proposed by your colleagues from New Jersey, Senator Bradley and Senator Lautenberg, and in the House, by Congressmen Pallone and Sharp. Most of my statement will focus on this legislative initiative, but I believe it is important that I put this initiative and the issues it confronts in the context of the pipeline safety program that will play a major role in carrying it out, should it be adopted.

THE PIPELINE SAFETY PROGRAM

Mr. Chairman, the Department's pipeline safety program has been in a constant state of flux for almost 10 years. With a growing list of mandates for action from Congress and the National Transportation Safety Board, most of which have followed accidents of national significance, and with resources incapable of keeping up with those mandates, the program has had to be one of crisis management rather than risk management. As such, we have established what I consider to be an admirable record for responding to accidents, although I believe we don't yet have a program that adequately addresses prevention.

In the last year, we have witnessed accidents that raise fundamental issues concerning the state of the pipeline infrastructure in this country and the need to better understand the risks posed to, and by, that infrastructure. We have been hard at work examining how best to manage these risks. I am pleased to report that, for the first time in the history of the pipeline safety program, I believe that the program will be equal to the challenges it faces. While this is partly a function of the promise of more resources, I believe as important is the personal attention of Secretary Peña to those challenges. In my experience, no previous Secretary of Transportation has been so well informed of the issues involved in assuring the integrity and viability of one of the Nation's largest transportation infrastructures.

By committing the Department, in a time of severe budget constraints, to pursue a budget for pipeline safety in fiscal year 1995 that is double what it was in fiscal year 1993, the Secretary has signaled a new era for pipeline safety. This will be an era in which the pipeline safety program can become a leader in risk management, and finally fulfill the essence of what its core purpose should be—a credible, technically solid, and independent assessor of the nation's pipeline infrastructure and actions of the industry to protect and operate that infrastructure. I believe that the proposed legislation can be an effective tool to help us realize that purpose.

NATIONAL ONE-CALL CAMPAIGN

Mr. Chairman, before discussing the proposed legislation, I would like to bring the Committee up to date on DOT's national campaign to improve the use of one-call notification systems. Over the last several months, we have taken significant steps in defining and implementing a strategy to strengthen one-call systems nationwide. DOT was an active participant in the annual One-Call Systems and Damage Prevention Symposium last May in St. Paul, and devoted an entire morning of the most recent meeting of our two Technical Advisory Committees to one-call issues. In addition we are:

- Examining reasons why participation in, and use of, one-call systems is not universal and exploring ways to enhance such participation and use.
- Compiling information on model state programs and best practices to share with other states which are developing their own programs.
- Working with various pipeline safety stakeholders to coordinate our efforts to improve one-call systems.
- Supporting state efforts to develop or strengthen their one-call laws.

In addition, the House and Senate Appropriations Committees are considering, in conference, funding of up to $1.5M which would allow us to make grants to states to enhance their one-call programs. We are very supportive of this effort and have identified several possible initiatives DOT could fund with that money including: education of excavators in safe digging techniques, education of the people who locate the underground facilities, development of software for use by states to easily comply with reporting requirements, and funding start-up costs of enforcement of one-call laws (many states do not have the staff initially to enforce one-call laws; once the system is up and running, enforcement can be funded out of user fees paid by participating members). Any or all of these projects would significantly improve the effectiveness of the pipeline safety program.

As you can see, DOT has been very active on the one-call front, and we will continue to vigorously promote and support increased one-call usage.

PROPOSED LEGISLATION

Mr. Chairman, you have asked for our views on proposals contained in S. 2101, "The Comprehensive One-Call Notification Act of 1994". Our comments are based on the latest version of the companion bill, H.R. 4394, (an Amendment in the Nature of a Substitute to H.R. 4394; Discussion Draft dated July 14, 1994).

The bill addresses the single leading cause of pipeline failures, external force, or third party damage, typically caused by heavy construction equipment striking a pipeline. This damage may be acute, that is, it causes the pipeline to fail immediately and puts construction crews and surrounding homes and businesses immediately at risk; or, it may be chronic, that is, it causes damage to the pipeline that over time results in failure, thereby putting the nearby population and/or structures at risk. It is this chronic damage that is proving to be the more serious and difficult to control.

The bill is predicated on-several concepts that we fully support, including:

- Comprehensive state-wide one-call notification systems.
- A federal pipeline safety grant program that provides funds to states to encourage state adoption of effective one-call programs.
- The essential elements of an effective one-call system: mandatory participation by all underground facility operators; only narrowly drawn exceptions to a general requirement that all excavators call before they dig; effective, easily administered sanctions; and, 24-hour coverage with provision for emergency excavation.
- Development by the Department of Transportation of a model one-call program.
- Development by the Department, through partnerships, of a nationwide public education program.
- An enforcement provision that leaves discretion to the states in establishing penalties and provides for increased penalties when damage results from a violation.
- Secretarial review of the adequacy of state one-call programs.
- Federal implementation of a nationwide toll free one-call telephone number.

However, we believe that several concepts included in the bill need clarification to ensure that they accomplish their objective. These include:

- The bill's requirement to establish a nationwide toll-free one-call telephone number. While supporting the concept, we believe that the requirement for a national number should not preclude the use of statewide numbers; rather the national number should be used in conjunction with statewide numbers. Many states have already invested significantly in publicizing their statewide numbers and eliminating them in favor of a national number might create confusion. The most beneficial use of a national number would be if an excavator is operating outside its normal base of operations (i.e., in another state) and calls a national number to be routed automatically to the appropriate number in the state where the excavation will take place. We think the language in the bill needs to be clarified to assure this outcome.
- The bill's provision for incentive grants to the states to establish new, or to improve, existing one-call systems. We support this but we need to work with the committees to ensure that it works as we believe was intended, i.e., the funds for these grants would be drawn from the existing balance in the Pipeline Safety Fund. The bill as currently drafted appears to draw these funds from user fee collections in the years for which the grants would be made. We believe the funds should be drawn from already collected but unexpended pipeline safety user fees.
- The bill's requirement for the Secretary of Transportation to develop a model one-call program. We support the concept of a model program, but we believe that the requirement should be combined with the statutory elements of the state program set out in Section 5 of the bill. This would simplify the legislation and ensure that two sets of statutory requirements (i.e., state program and model program) do not inadvertently confuse the desired result.

There are a few additional elements in the legislation which we believe need further work, notably the compliance provisions in Section 4(c). We have detailed our concerns in an attachment to this testimony. We will continue to work with the committee staffs to resolve these concerns.

CONCLUSION

Mr. Chairman, the pipeline safety program is on the threshold of becoming what it has needed to be for many years. The Secretary has pledged his personal support to assure that the program reaches its full potential. I know I speak for our political leadership when I add our pledge for the same outcome. To reach our full potential will take more than resources—it will take a new leadership approach that seeks out partnerships.with states, local governments, public interest groups, and the industry.

The successful management of the risks inherent in pipeline transportation will require not only sound public policy decisions by all levels of government, but will require a wider and deeper understanding of pipelines—where they are, what they do, how they are maintained, and what risks they pose. If we are successful in nothing else in the "new Office of Pipeline Safety," we must succeed in increasing public awareness of pipelines—and assuring that such awareness occurs before an accident. We believe that Congress has made great progress towards crafting one-call legislation that will significantly enhance our ability to meet this objective. We will continue working closely with the committees to produce an effective bill.

Thank you.

REMAINING DOT CONCERNS WITH S. 2101 "COMPREHENSIVE ONE-CALL NOTIFICATION ACT OF 1994"—AUGUST 3, 1994

- Section 4, paragraph (c), Compliance

This provision gives the Secretary or any aggrieved person a cause of action against a state if the state fails to comply with he law. We are concerned that such a freestanding mandate to the states raises substantial Tenth Amendment concerns.

We need to have further discussions with the committee staffs to ensure that any compliance provision is meaningful and clear.

- Section 6, paragraphs (b) and (e), Penalties and Enforcement

These provisions providing that the state "may * * * assess" a penalty of imprisonment and "may include" on-site citations, are superfluous since they merely reflect what a state already has the power to do.

- Section 7, paragraph (a), subparagraph (1), Coordination

This provision requires the Secretary to coordinate implementation of this Act with the pipeline safety laws. It is not clear from the language if this is just a routine request to ensure there are no conflicts among the laws, or if there is some special significance that would impact how this Act is implemented (e.g., would it have an effect on which pipeline facilities would be included under this bill?).

Senator EXON. Mr. DeLeon, thank you very much. Mr. Batten, you were not on the witness list, but if you have anything to say—I understand you are here as an associate of Mr. Hall; is that correct?

Mr. BATTEN. That is correct, sir.

Senator EXON. Gentlemen, I have no basic questions for you. We may have some questions for the record. I appreciate your being here. I appreciate that you have come forward as the people who have to deal on a day-to-day basis with these tragedies. Your support for the program is very, very important and is well noted. Suffice it to say, we have to do more in this area than we have done in the past.

Since you were good enough to mention the situation with Amtrak, it emphasizes the importance of you safety people and the examinations you do after the fact. The examination you do after these tragedies occur is very important in helping us understand what we need to do legislatively to try and prevent some of these incidents. We still are going to have accidents. You are here alerting us how you see the problem so we may move forward on this legislation in hopes of preventing at least some of these accidents.

Please feel free to give us any additional input by letter, telephone calls, or direct communications as we move forward on this bill.

With that, I thank you for being here and you are excused.

Mr. DELEON. Thank you, Senator.

Mr. HALL. Thank you, Senator.

Senator EXON. Panel 2 consists of five individuals whom we are very pleased and honored to have with us today. Suzette Denslow, deputy director, Virginia Municipal League, representing the National League of Cities; David McNiel, vice president, technical services, Tenneco Gas Pipeline Group, representing the American Gas Association and the Interstate Natural Gas Association of America; William P. Boswell, vice president and general counsel, the Peoples Natural Gas Co., representing One-Call Systems International; Larry Thomas, president, BP Oil Pipeline Co., representing the Association of Oil Pipelines; and Paul E. Devaney, member of the technical staff, Bellcore Corp., as chairman, Utility Location and Coordination Council, American Public Works.

We certainly welcome all of you here. I see, Mr. Devaney, that you have a New Jersey address, and possibly that has something to do with your keen interest in this hearing. You may be able to shed some additional light on do's and don'ts in the legislation which was introduced in response to that tragedy in New Jersey.

Let us begin with Ms. Denslow. Thank you for being here.

STATEMENT OF SUZETTE DENSLOW, DEPUTY DIRECTOR, VIRGINIA MUNICIPAL LEAGUE, REPRESENTING THE NATIONAL LEAGUE OF CITIES

Ms. DENSLOW. Thank you, Mr. Chairman, for the opportunity and I appreciate all of the work you have already done on this issue.

I would like to address two main points today. One is the need for a Federal model program and the other is the issue of administrative and civil penalties.

I thought there were only four States that did not have the system, but I hear today only one State does not. I would say that a model program would be useful if it focuses on what the goals are of the Federal legislation. I think we would be less excited to see a program that has a lot of implementation specifics that would make States change their existing programs, and many of the existing ones are good.

On administrative and civil penalties, first I would say in general that if you have penalties that are too high you are going to deter people from reporting damage and we would discourage that. We would rather know where the damage is, especially if it is just structural weakness that is caused, and fix it rather than impose high fines.

Second, and more importantly to local governments, is the issue of administrative and civil fines on local governments, and we would discourage that. We think that fundamentally local government utilities are different from other utilities. We operate a variety of utilities, water, wastewater, gas, electric. We own and operate on the land that we own. And another primary function of local governments is public protection, so when we break something we are also in charge of fixing it, cleaning it up, and overseeing all the protection.

We find in Virginia less than 5 percent of the damages are caused by local government utilities. We feel we are not the problem.

If I could take one second to talk about our city of Richmond, to give you some concept of the size of the system, the city of Richmond has 1,300 miles of gas main, 1,200 of water main, 1,200 miles of sewer main. The water system is over 100 years old, 160-some years old.

Now, when you talk about the issue of mapping, and Senator Lautenberg talked about mapping our entire Nation, older cities cannot map where all their water and sewer lines are if they are old. It does not matter if a water line is here or here if it is 100 feet underground, and that is what we have got. No one is digging that far to hit it, and so the mapping would be a very costly thing for us to do. New lines are not a problem.

Richmond already, because we have a mandatory system in Virginia, responds to 35,000 calls a year from the one-call legislation. So, I say that because, one, we have a pretty big utilities operation at the local level, and two, we are already doing a lot in this area.

In terms of why we do not feel that we should have punitive administrative or civil fines, the answer is, one, that the purpose is to punish or to prevent. A fine on a local government punishes residents not the local utilities. And two, the best penalty for elected officials is public embarrassment and that is what happens when we have problems and we break lines is there is public embarrassment, so we would rather see that.

My specific suggestion is that on H.R. 4394, the July 14 edition, in the general penalties section you add a line that says "provided, however, that each State may exempt local governments from all administrative and civil penalties." And that would pretty much take care of our problems with the bill.

Mr. Chairman, that concludes my remarks. I would be happy to answer any questions.

[The prepared statement of Ms. Denslow follows:]

PREPARED STATEMENT OF SUZETTE DENSLOW

Senator Hollings and members of the Senate Committee on Commerce, Science, and Transportation, I am Suzette Denslow, Deputy Director of the Virginia Municipal League. I very much appreciate your invitation to testify on the Comprehensive One-call Notification Act of 1994. This is the "Call before you dig" program that makes it easy for homeowners and excavators to get quick, reliable information on the location of underground utility lines.

First, much work has been done in developing S. 2101 and in working out the details of how a one-call system should operate. We greatly appreciate the efforts you have made to listen to the concerns of the utilities in general, especially the local government utilities. In addition, much work has been done on the Amendment in the Nature of a Substitute to H.R. 4394. The house bill at this point incorporates many of the changes proposed by local government officials and other utility operators. Rather than walk through the point-by-point differences between the Senate and Mouse versions as they currently stand, I would prefer to focus on H.R. 4394.

I will address my concerns with proposed legislation on two fronts: the issue of the need for a Federal model program, and the issue of administrative and civil penalties.

NEED FOR A FEDERAL MODEL PROGRAM

Most states already have one-call legislation. A Federal model program may be a good outline for a state about to start up a new program, but there are only four states that do not have a one-call system in place. In Virginia, and in many other states, we have very good systems that are working effectively, and we have reservations about a Federal model that would change our existing systems. I am sure that the experience in each state is the same as that in Virginia: our one-call legislation was the result of years of work and reflects the unique needs in Virginia. H.R. 4394 does include provisions for "self-certification" of state systems (for compliance with the Federal model), and it puts the onus of disapproval on DOT. I think that the onus on DOT should be to prove that a state's system is not in keeping with the goals of the Federal law, rather than specific which will be determined at a later date. All utilities agree with the goal of the Federal law, to protect the public from the repercussions of damaged underground utilities. On behalf of the National League of Cities and some of the more rural states, let me repeat what the Montana League of Cities and Towns urged: "[please don't] pre-empt state laws that are based on [local] experience, financial necessity and common sense."

ADMINISTRATIVE AND CIVIL PENALTIES

First, I would discourage the use of very high fines in general. If you have a contractor who hits a line but does not break it, he may cover it back up and not report the 4amage rather than face a $25,000 fine. Even though the line didn't break, there may be a structural weakening from the hit that could be repaired easily. If the repair is not repaired early, the weakening can create a break later. We would prefer that damage incidents all be reported. The fining procedure should not be a deterrent to reporting.

On the specific issue of municipalities, let me start with the concept that local governments are a different breed of utility owners. We serve as the owners, operators, contractors and excavators for the several varieties of utility systems: gas, water, wastewater, electric, storm sewer systems, etc. A fundamental difference between government and other utilities is that we own the property on which we operate, construct, and excavate for maintenance and repair. Generally, other contractors and excavators are using public right-of-way for most of their work—we are allowing them on public land when they locate and work on their lines. Each of the aspects of utility operations are our day-to-day business: locating our lines for others, providing water and power, removing sewage and stormwater, and constructing, repairing and maintaining lines.

Further, public protection is local governments' day-to-day business. When a line breaks, local government can't just pay a fine or penalty and walk away. Local government is the "point person" from start to finish of whatever clean up, repair, and restoration of service takes place. It is our distinct disadvantage for breaks to occur, whether it is caused by our employees or others. All of this is borne out by a 1994 study of Virginia's Underground Utility Damage Prevention Act, which shows that less than 5 percent of damage incidents were caused by local governments. Homeowners were the cause of twice as many damage incidents as all local government utilities. Sixty percent of the damage incidents were caused by non-government utility and other contractors. We—local governments—are not the problem.

Let me cite what we are doing already by using the specific example of one of Virginia's municipalities, the City of Richmond. Richmond has 1,300 miles of gas mains and 1,200 miles each of water and sewer mains. Richmond's water system was started in 1830 with wooden mains—none of which are in use any more to the best of our knowledge. Most of the water mains are old. The problem with old systems is that you don't always know the location of the old lines, and many of the old water and sewer lines are very deep. Old lines can be as deep as 100 and 200 feet. Of the 1,300 miles of gas lines about 600 miles are old cast iron mains which the City stopped installing in the 1960's. This, too, is an old system. Richmond, like any old city has the challenge of operating, maintaining and adding on to a system that has been in place for more than a hundred years—in Richmond's case, 164 years. In addition to the normal utilities operations, Richmond also must locate its lines for other excavators. Currently, the City of Richmond receives about 35,000 calls a year through the one-call system to locate gas and water lines—this amounts to about 135 calls per work-day. The city operates three crews all the time just to respond to the location calls.

I say this not only to point out the magnitude of a city system and some of its inherent problems, but also to point out that we are already doing a lot in this area. We are already meeting the goals of the proposed legislation.

In Virginia, for several years, local governments have been required by state law to participate in the one-call system that we call "Miss Utility." We are full participants and we follow all the standards that the non-government utilities follow. When the law was strengthened this year, we supported the strengthened procedures. In addition, all of the state's utilities, government and non-government, supported exemptions for municipalities from administrative and civil penalties. We felt this was appropriate for our local governments for several reasons. First, the purpose of punitive damages is to punish. Any fines levied against a local government will be paid by the residents of the locality. The government entity per se is not punished by fines, the residents are. We are not in the business of making profits, so fining us does not hurt our profits. Second, the punitive damage that is appropriate for local governments is the one that is already used and will continue to be used: public embarrassment. As elected officials, you can appreciate that public notice and newspaper reports and public outcry are all tools used to get results from government. These tools have more important effects on government operations than punitive or administrative fines.

My specific recommendation on H.R. 4394 is that an amendment be made to "Sec. 6(a) General Penalties," to add the following phrase to the end of line 18: "provided, however, that each state program may exempt local governments from all administrative and civil penalties under this section." This amendment simply would allow each state to determine for itself how to treat its local government utilities. There would be no requirement for exemption or inclusion, but there would be a state option for either.

Mr. Chairman, that concludes my remarks. Again, I appreciate the opportunity to appear before you today, and I would be happy to answer any questions.

LETTER FROM CHRIS MCKENZIE, EXECUTIVE DIRECTOR, LEAGUE OF KANSAS MUNICIPALITIES

JULY 29, 1994.

CARA WOODSON and FRANK SHAFROTH
National League of Cities
Washington, DC 20004

Thanks very much for your memo of July 22, 1994 in which you requested comments on the proposed "Comprehensive One-Call Notification Act 1994". My comments are as follows:

1. Since Kansas already has an underground utility damage prevention act (see enclosed article from Kansas Government Journal), there is little, if any need for such legislation in the State of Kansas. While Judy O'Brien's memorandum states that the bill makes it clear that if the state does not have to run a program or establish a brand new program if one is already in placed, Section 4(a) contains an ambiguous mandate for "each" state to establish a comprehensive statewide one-call notification program. This requirement can be met, according to Section 4(a), "by modification or improvement of an existing program," but one has to question what "modifications" or "improvements" would be necessary in order to comply.
2. Section 3 establishes a new nationwide toll free number system for One-Call purposes. I suspect that this will do nothing more than confuse local officials who are already getting used to using the state One-Call system.
3. The definition of "underground facility" includes water and sewer facilities. The Kansas Underground Utility Damage Prevention Act excludes water and sewer operations from its definition of "underground facility." This exemption of water and sewer facilities has made implementation of the One-Call notification mandate much more feasible and acceptable to Kansas local officials, and I would urge a similar exemption in any federal legislation.

Thanks again for the request for comments. As you can probably tell, I am not convinced this legislation is needed at all. I appreciate your efforts to make it more workable for the local officials across the country.

LETTER FROM KENNETH G. BUECHE, EXECUTIVE DIRECTOR, COLORADO MUNICIPAL LEAGUE

JULY 25, 1994.

CARA WOODSEN
National League of Cities
Washington, DC 20004

The short turn around required by your July 22 fax has allowed us to provide only a brief, preliminary review of the July 14 draft. The following comments are offered under these time constraints:

- We recommend that NLC oppose the legislation as an unnecessary and inappropriate federal intrusion into state and local operations. The vast majority of states already have state laws on the subject.
- The "model program" set forth in this bill tends to conceal the mandatory requirements of the bill. Once the rules and regulations are issued and the bill's requirements are implemented, another intrusive federal mandate will be upon public and private entities.
- The costs of complying with the requirements are not just state government compliance costs. What about the notification and compliance costs for local govern-

ments, property owners and excavators? How are local governments going to recoup their costs?

• Let's not superimpose federal liability requirements on existing state liability laws.

If we develop more specific comments later but in a timely fashion, we will send them to you.

Thank you for sharing the draft with us.

Sincerely,

KENNETH G. BUECHE,
Executive Director.

LETTER FROM JOHN P. COLEMAN, EXECUTIVE DIRECTOR, OHIO LEAGUE OF CITIES AND TOWNS

JULY 22, 1994.

FRANK SCHRAFROTH
National League of Cities
Washington, DC 20004

Ohio has an Underground Utility Protection Service. The League worked with the state and public utilities to determine what mandates would be imposed on municipalities.

Municipalitiee are required to include in all construction plans, the location of underground facilities. Municipalities must contact the Underground Utility Protection Services and the owners of underground facilities which are not members of the service. The contractor, two days before digging, must contact the owners of the underground facilities.

If the municipality does not comply with the law, the municipality is liable for damages which may occur.

The program is working and we oppose any mandated federal model program. Elements included in the federal model must include fees necessary to operate the program and administrative or civil penalties. Municipalities may use the one-call Ohio service but are not required to be members of the service.

LETTER FROM ALEC HANSEN, EXECUTIVE DIRECTOR, MONTANA LEAGUE OF CITIES AND TOWNS

JULY 22, 1994.

FRANK SCHRAFROTH
National League of Cities
Washington, DC 20004

DEAR FRANK: Montana law, copy enclosed, accommodated rural areas where a one-call system world be financially out of reach. In our state, utilities located in areas that do not have a one-call system, file the name, phone number and address of the appropriate contact person with the County Clerk and Recorder. Excavators are then required to make all the contacts on the list before they begin work.

If Congress Is going to get into this business, it must not preempt state laws that are based on experience, financial necessity and common sense. The Montana League would be opposed to any federally mandated one-call system that goes beyond what we are reasonably able to provide in rural areas.

Thank you for contacting me about this issue.

Best wishes,

ALEC HANSEN,
Executive Director.

Senator EXON. Thank you very much, Ms. Denslow. I would like to ask you at this time about old pipelines to which you referred earlier. Evidently, there was no mapping of those when they were laid; is that correct?

Ms. DENSLOW. There may have been some scant mapping, but pretty much at the time—my impression is in the 1830's if you laid a water main you knew where it was, and so they were not thinking about the future of 1,300 miles of water main.

We know about where it is because we know it is going from one building to, you know, a main out on a street.

Senator EXON. But you could not draw it on a map close enough to qualify under one-call; is that what you are saying?

Ms. DENSLOW. In all cases. For a lot of the lines we absolutely would not have a problem, but for some of the older lines unless you are building, excavating for a new building, you are not going to go down deep enough to get to them.

So, we would say no mapping them. We do not have existing maps. And for any older city in the Nation I think you would see the same case.

Senator EXON. I would ask you and the rest of the witnesses who may have any information on devices which we have or do not have that could identify these things. My wife has one of those gadgets she uses to find pennies on the ground and those kinds of things. I never understood how they worked but they are pretty ingenious. Has there been no development of a similar technology that would go down 1, 2, 10, 20, or 30 feet and identify either telephone lines or pipelines or water mains?

Ms. DENSLOW. Mr. Chairman, I probably know less than your wife does. I would defer to one of the other members.

Senator EXON. Thank you very much. Let us proceed on down the line, then. I am sure that there is no easy solution to this problem or it would have been done before. I just keep wondering whether or not, with all the funding we do for new technologies, if this is something that we could hurry along if private industry has not responded in this area. If it is workable, I am surprised they have not. Maybe it is not workable.

Mr. McNiel, I am pleased to recognize you and welcome. I first would like to say I notice you are with Tenneco. I am just curious to know, on a personal note, whether you were close to or had a chance to meet my good friend Mike Walsh, as the head of your organization, before he passed away.

Mr. MCNIEL. Yes, sir. Mike Walsh and I met on a number of occasions. He was a great leader for our company and we were sad to see his passing.

Senator EXON. Thank you very much. Please proceed at this time.

STATEMENT OF DAVID McNIEL, VICE PRESIDENT, TECHNICAL SERVICES, TENNECO GAS PIPELINE GROUP, REPRESENTING THE AMERICAN GAS ASSOCIATION AND THE INTERSTATE NATURAL GAS ASSOCIATION OF AMERICA

Mr. MCNIEL. Mr. Chairman, I am David McNiel, vice president of technical services for Tenneco Gas headquartered in Houston, TX, and I am pleased to testify today on behalf of both the Interstate Natural Gas Association of America, or INGAA, and the American Gas Association, the AGA, concerning Senate bill 2101, which was introduced by Senator Bradley and which concerns a nationwide one-call system.

Before I get started, we would like to thank you, Mr. Chairman, and Chairman Hollings, as well as Senator Bradley and Senator Lautenberg, Representative Pallone and Chairman Sharp, and all of your staffs for your support on what we believe is a critical issue for our industry and for our country.

Both INGAA and AGA strongly encourage Congress to pass one-call legislation this year. That one-call system must provide excavators with the convenience of being able to place a single call to a one-call center, and then being able to have that center dispatch the operators of all underground facilities in the vicinity of that excavation to their site, and mark that pipeline or TV cable or telephone lines prior to the beginning of the excavation. That is the essence of a one-call system.

We are fully aware that, today, Federal, State, and local governments all have many different types of needs, each of which are competing for limited funds. And as you have heard already this morning, Mr. Chairman, one of the most important things to note about a one-call legislation, particularly as it has been adopted by many States today, is that those facilities and those one-call centers are funded primarily by the operators of the underground utilities themselves, and often funding is in concert with the excavators or contractors themselves. I think that this approach addresses one of the key issues that you raised earlier this morning.

Mr. DeLeon referenced a number of the accidents that involve third-party damage. In fact, since 1970 in excess of 60 percent of all pipeline incidents involving transmission and distribution systems were the result of third-party damage, and more seriously than that, almost one-half of the fatalities in that same time period were the result of pipeline incidents caused by damage from third parties.

Unfortunately, very few of the State programs require all of the excavators and all of the underground facility operators to participate in their programs. Even more importantly, few of those programs provide for statewide coverage and in a number of cases there is not an enforcement program with real teeth in it.

One-call legislation should contain certain basic elements in our opinion. First, there should be a single, easily remembered nationwide telephone number that all excavators, all entities, can call. Of course the benefit of a single, nationwide number akin to a 911 or a 411 for information about directory listings is that you can use that number across the country in an education program to really help the public become more and more aware, and the excavators become more and more aware, of the need to call before they excavate near other facilities.

Second, we believe that the Federal Government should review and revise the minimum standards that have already been set for the establishment of one-call systems, and then within that framework permit the States great latitude, or as much latitude as they need, in order to implement an effective damage prevention program.

The Department of Transportation should install a model program that should ideally have four key elements in it. First, and we mentioned this earlier in previous testimony, all operators of underground facilities must be members of that organization or the one-call center. Participation is a key.

81-903 O - 95 - 3

Second, all excavators who are undertaking activities that could result in a striking of a pipeline or any underground utility must call before they dig.

There are certain situations in which the excavator and the operator of the underground facility can be one and the same. In those cases where there is certainty that there are no other underground utilities in the area, then perhaps a call to the one-call center would not be necessary. But that should be a very small exception to the rule.

Third, enforcement should accomplish several objectives, Mr. Chairman. It should, one, encourage contractor participation. It should punish failure to make the call, and it should strongly encourage the excavators who do in fact strike the line or strike the other underground utilities to make that call promptly so that the operators of those facilities can get out and make the appropriate repairs prior to any significant event occurring.

But within that context we need to make sure that the States' programs are not so punitive that excavators are fearful of reporting that they have accidentally struck a facility. It is very, very important, as you have heard before, that we know that our facilities have been hit. That allows the operator to go out and take care of the damage before it becomes a significant problem.

And then, fourth, the Department of Transportation model program should include a public education component which the Department of Transportation would complement with its own effort.

We are pleased to see that these concepts have been included in Senator Bradley's legislation. We understand that there have been some concerns expressed and we, along with all parties, agree that consensus language must be adopted in order to pass a bill this year, given the time remaining in Congress.

So, in closing, Mr. Chairman, I want to promise you and the other members of this committee the continuing assistance of our industry in passing solid, one-call legislation which the natural gas industry and the public need in order to protect the underground facility infrastructure in this country.

Thank you.

[The prepared statement of Mr. McNiel follows:]

PREPARED STATEMENT OF DAVID MCNIEL

Good morning, I am David McNiel, Vice President, Technical Services, Tenneco Gas, headquartered in Houston, Texas. I am pleased to testify today on behalf of the interstate Natural Gas Association of America (INGAA) and the American Gas Association (A.G.A.) on a bill, S. 2101, introduced by Senator Bradley dealing with nationwide one-call systems.

INGAAA is a non-profit national trade association representing the interstate natural gas transmission companies operating in the United States and Mexico, and the interprovincial pipelines in Canada. A.G.A. is a national trade association comprising approximately 275 natural gas distribution and transmission companies throughout the United States, Canada and Mexico. Together our member companies transport and deliver 90 percent of the nation's natural gas from the wellhead to the burnertip through the nation's 280,000 miles of interstate transmission pipelines and 1.5 million miles of distribution pipelines. Our members serve over 56 million customers in the United States.

INGAAA and A.G.A. encourage Congress to pass one-call legislation this year. One-call systems should provide excavators the convenience of placing a single call to a one-call center to have all underground facilities in the vicinity located by the facility operators before digging begins.

We wish to thank you Mr. Chairman, as well as Chairman Hollings, Senator Bradley, Senator Lautenberg, Representative Pallone, Chairman Sharp and your staffs for your support of this most important issue.

On behalf of INGAA and A.G.A., I would like to also express our appreciation for the excellent work of the Utility Location and Coordinating Council of the American Public Works Association and also the Department of Transportation's (DOT) Office of Pipeline Safety (OPS) who have labored so diligently over the years to develop one-call systems throughout the nation. We urge Congress to recognize these efforts and to carefully build on the progress that has already been achieved.

The need for one-call systems arose as more and more utility companies—gas, electric, cable, etc.—have been placing their facilities underground, especially in populated areas. These areas are also frequently high growth areas and therefore experience a great deal of construction activity and associated excavation. In the 1970's the first "one-call" centers were established to protect both the excavator and the underground facility.

Much progress has been made since that time. Forty-six states and the District of Columbia have enacted one-call legislation that basically supports a one-call notification process. Twenty-seven states make one-call mandatory. However, improvements are still needed In many areas. According to DOT data, outside forces or third party damage has caused over 63 percent of the natural gas transmission and distribution incidents reported since 1970 and nearly one-half of the fatalities. Unfortunately, few states require all excavators and all underground facility operators to participate, nor do they provide for statewide coverage or have an effective enforcement program.

In this age of competing needs for limited funds at the state and local level, it is important to note that one-call programs are funded by the participating underground facility operators, sometimes in conjunction with dues paid by excavators. Charges are set by negotiation with the one-call center operator and the underground facility operators and excavators. Facility operators pay the charges usually on the basis of the number of calls they get from the one-call center that require a locate. There is usually no charge if no locate is required.

We recommend that any federal comprehensive one-call legislation be carefully crafted to avoid creating mandated federal or state operated one-call systems, but rather provide guidance, direction and incentives to the states to improve existing systems or create statewide systems. This legislation should contain certain basic elements to encourage and reward states for developing and implementing comprehensive effective one-call systems.

First, there should be a single easily remembered nationwide telephone number dedicated to reporting planned excavations. Most reports will originate in the same area code as the excavation so that simply dialing this number could connect local excavators with their one-call center. Where excavation is planned in a different area, prefixing the dedicated number with the area code for the excavation location could reach the proper center, Use of this number should be free to excavators so as to encourage reporting. The benefit of a single nationwide dedicated number is that it can easily be popularized through a nationwide public information campaign. This might form the center focus of a national effort by underground facility operators to bring excavation safety to the public's attention. Expertise in the Federal Communication Commission will be very important in establishing how this communication system will operate.

Second, the Federal Government should review and revise the minimum standards they have set for the establishment of one-call systems and permit states flexibility in meeting these standards. The Department of Transportation should establish a model that could be copied by states to create effective damage prevention programs. Ideally, such a model should include the following:

a) All operators of underground facilities that are at risk of being struck by outside excavators should be included in the program. Historically, underground facility operators have found that it is more cost effective to participate in one-call systems to prevent damage rather than having to repair facilities after the fact.

b) All excavators undertaking activities that could result in the striking of underground facilities should be required to first call the one-call system (unless the excavator and the facility operator are one in the same person and no other underground facilities are threatened).

c) Enforcement should encourage contractor participation, punish failure to call and strongly encourage excavators who strike facilities to promptly report such damage. States should tailor their programs such that excavators are not fearful of reporting when they accidentally strike facilities.

d) The model should include a public education component which DOT would complement with its own effort.

e) DOT should periodically update its model program.

DOT should report on whether state programs meet the model federal guidelines, publicize the results of the report and recommend actions that should be taken to encourage states to upgrade their programs. DOT should also hold workshops for states to educate them on the benefits of good one-call systems.

We are pleased to see that these concepts are included in Senator Bradley's legislation. Both the INGAA and the A.G.A. boards of directors strongly support passage of one-call legislation embodying the above principles. We believe that it is very important to enact legislation this year. Some organizations have expressed concerns with the language of S. 2101 and H.R. 4394 as introduced. Chairman Sharp's staff has been diligently working with all affected parties to develop consensus one-call language. Chairman Sharp will be holding a mark-up of H.R. 4394 tomorrow, August 4th, in his subcommittee. INGAA and A.G.A. have reviewed his substitute amendment to H.R. 4394 and are in support of its adoption. At the same time, all parties agree that consensus language must be adopted in order to pass a bill this year given the time remaining in Congress.

In closing, I want to promise you, Mr. Chairman, the continuing assistance of our industry in passing solid one-call legislation which the natural gas industry and the public need to protect the underground service infrastructure.

Senator EXON. Thank you very much, Mr. McNiel. Mr. Boswell.

STATEMENT OF WILLIAM P. BOSWELL, VICE PRESIDENT AND GENERAL COUNSEL, THE PEOPLES NATURAL GAS CO., LEGAL ADVISER, ONE-CALL SYSTEMS INTERNATIONAL; ACCOMPANIED BY WILLIAM KIGER, EXECUTIVE DIRECTOR PENNSYLVANIA ONE-CALL SYSTEM, INC.

Mr. BOSWELL. Mr. Chairman, good morning. I am William P. Boswell, and with me is William G. Kiger. We have previously submitted a statement, which I ask be admitted into the record.

We represent OCSI, One-Call Systems International, which for 20 years has been the umbrella organization for one-call information training and education efforts throughout North America. OCSI is a part of the American Public Works Association.

The one-call concept itself has been around for about 30 years, although most of our efforts have occurred since the mid-1970's. Currently, there are one-call systems in every State except Hawaii, and the systems take some 30 million calls each year and turn them into some 300 million messages to operators of underground facilities regarding proposed excavations.

If we leave you with no other message, it is that one-call is working, and it is working well. It can be made better. We concur strongly that legislation be passed this year, so I echo my previous colleagues here that that would be a very good thing which we think would enhance the program.

We identify four key principles for effective operation of the one-call system. They are: mandatory membership, a duty to call, continuing education, and a graded enforcement system. All owners or operators of underground facilities serving the public must be members of a one-call system. This applies, as well, to the State departments of transportation and to municipalities, which frequently have been exempted.

All excavators using mechanized equipment must be required to call a reasonable time in advance of excavation, except in the case of an emergency, in order to permit facility operators to have sufficient time to mark the locations of those facilities. By the same token, they cannot call too far in advance, or circumstances may change in the interim.

Excavators must continually be educated in the need and benefits of calling. They need to know how to do it, why to do it, and why it is in their interest and the public's interest that it be done.

Penalties for not calling are necessary, but only as a last resort, and then only in relation to the offense. Administrative sanctions generally are the most effective, and more effective than civil and criminal penalties because of the ease of administration.

Getting people to call is the most important consideration. Getting people who have not called to call if they hit a facility is the second most important consideration.

Up to this point, the main action regarding one-call on underground protection has been at the State level. The Federal Government is a relative newcomer, which is an advantage in one respect because it can benefit from the education we have received at the State level over the past 30 years.

Penalties for noncompliance will be necessary in some cases, but they should never be so large nor so inflexibly applied as to discourage people from calling. An excavator who hits a line should not be so terrified to report the hit that he covers it up and hopes for the best. The proposal that the States have a market basket of penalty provisions suitable to the offense is appropriate. The penalties should move from administrative sanctions to civil fines and criminal sanctions.

Of these, administrative penalties—for example, a citation for, say, $250, suspension of bidding rights—are the most effective because they are swift, easy to administer, and they get the point across.

The Federal Government can play an important role in the process. First, it can recognize that much of what happens has and will continue to happen at the State level. Thus, it should encourage the States to find ways of improving the damage prevention programs already in place.

It likewise should encourage the States to have systems which reach broad, uniform goals. These previously have been listed, but to repeat, they are: mandatory membership, a duty to call, continuing education, and graded enforcement.

Through the designated agency, which I gather will be DOT, the Federal Government ought to work with interested parties to produce a model one-call damage prevention program, and the sections of the July 14 draft which address this are excellent.

The Federal authorities should assist the States and the one-call systems in developing pilot projects to improve the delivery of service, and the Federal Government should actively pursue public education as the best means of getting people to call and respond.

All in all, the July 14 working draft—and we have seen a later version of it—with some other minor modifications would produce an excellent Federal law, one which would recognize that damage prevention is a national issue cutting across State boundaries. It would permit the DOT to act as a facilitator and a focal point, yet it would leave it to the States to work out the particulars themselves. We believe this is a superior path in a Federal system.

Three points that came up in earlier testimony, Mr. Chairman: one-call is paid for by its members, and occasionally by contractors doing excavation, so Mr. Pallone is right on point there. That will not change under this law.

In Pennsylvania, for example, on the issue of municipalities, three-quarters of our municipal members are exempt from paying anything at all, and that is by State statute, and it works just fine.

In Pittsburgh, for example, which is a large city and does pay, they paid $4,300 in 1993 as a member of the Pennsylvania One-Call System. In return for that, their permit revenue from issuing excavation permits was in excess of $350,000, so it was a money-making proposition for the city of Pittsburgh, and that is pretty much the case where the thing is tied into the permit system.

Last point, you inquired about detectors of underground facilities. If it is metallic, it can probably be detected, but only horizontally, not vertically. There is nothing on the market that is any good that we are aware of which can tell you how deep a piece of pipe is. Maybe some of the downward-looking radar that the military has would work, but it may be prohibitively expensive, and the best we can do in most cases is give you where it is located horizontally, within certain limitations.

That concludes my testimony. I would be happy to answer any questions.

[The prepared statement of Mr. Boswell and Mr. Kiger follows:]

PREPARED STATEMENT OF WILLIAM P. BOSWELL AND WILLIAM G. KIGER

Good morning. I am William P. Boswell, and I appreciate the opportunity to present testimony prepared by William G. Kiger and myself regarding S. 2101 and the current markup of HR 4394, which, among other things, would have the States implement one-call systems as a part of their damage prevention programs for underground facilities. I previously testified on June 23, 1994 before the House Energy & Power Subcommittee on Mr. Pallone's companion bill.

By way of introduction, Mr. Kiger is executive director of the Pennsylvania One Call System (POCS), a Pennsylvania not-for-profit corporation, and chair of the legislative committee of One Call Systems International (OCSI). OCSI is a part of the American Public Works Association (APWA), and is the umbrella organization for educational and other purposes for the one-call centers throughout North America and those other areas of the world where they exist. He is a past chair of OCSI, and helped found the organization in 1975. He also serves as vice chairman of the Utility Location and Coordinating Council of the APWA.

I have been legal advisor to OCSI since 1980, and I am legal advisor emeritus and past chair of POCS. I am vice president and general counsel of The Peoples Natural Gas Company, Pittsburgh, Pennsylvania. Of some relevance on this subject, I also am an elected public official, serving as president of the borough council of Osborne, Pennsylvania. I have written, spoken, narrated videos, and otherwise been a key, volunteer player in the one-call effort throughout the country for more than 16 years.

The companion bills are a welcome step toward a constructive federal role in a process that primarily has been addressed at the State level for the past 30 years. We emphasize that one-call damage prevention programs are widespread and that they work exceedingly well. All but two States have some form of one-call system (this will drop to one State as of 1/1/95), and all but four states (three as of 1/1/95) have damage prevention statutes. Collectively, OCSI members received more than 20 million calls to our centers in 1993, which generated some 300 million work site notifications to our participating facility owners and operators. These messages, called-in and received, dramatically reduced the risk incidence of damage to persons and property. One-Call systems are constantly evolving and improving, but one thing is certain: the excavator must make the call, and the facility owner or orator must do something with the call it receives. Each of these actions is more likely to occur with proper education of the parties concerned.

The bills could help move damage prevention efforts toward greater predictability from State to State while continuing to give each State the flexibility it needs to experiment and to deal with local problems the most effective manner. We urge that any action by Congress recognize both the value of what already exists, and be in support of State one-call efforts, which have been underway for over a quarter of a century.

As a general proposition, we favor uniformity in certain specific areas.

First, all owners and operators of underground facilities serving the public should be required to be members of one-call systems, and to participate in damage prevention programs.

Second, those who excavate should be required to give advance, timely notice of their intention to do so, and they should be able to give such notice as easily as possible.

Third, there should be continuing education programs, the object of which would be to advise persons of the risks inherent in excavation, and of their responsibilities to minimize those risks by complying with the notice and other requirements of State law.

Fourth, where voluntary compliance does not work, there ought to be several means of enforcing notice and participation mandates in a timely and constructive fashion; these means should include administrative sanctions, injunctive relief, and civil and criminal penalties, in that order.

PARTICIPATION

Traditionally, the owners and operators of underground facilities, and the large contractor organizations, have been key supporters of and participants in one call and damage prevention programs. Much of this interest is related to the fact that it is their facilities, their equipment, and their employees who are at the greatest risk of damage and injury. Conversely, there are groups which traditionally have resisted participation; usually these have been the State departments of transportation and organizations representing local municipalities. It is ironic that these two groups are among the larger excavators and are themselves facility owners, with many underground facilities located within their public rights-of-way.

Often, the reasons cited by State DOTs for their reluctance to participate are as these: first, they claim they never cause any damage; second, they claim it would be too inconvenient to their work schedules if they had to call ahead of time; or third, they claim they already know where all of the underground facilities in their jurisdiction are located.

Unfortunately, empirical evidence shows that these assertions are not correct. Traditionally, State DOTs have been exempted by State statute, in many cases as a result of he threat of a gubernatorial veto of the legislation if they are included. As a half-way measure, some State DOTs have agreed to a requirement that outside contractors working for hem, but not their own (State) employees, be covered by the law and thus required to call. In our opinion, there is no logical reason for a state DOT to be excluded; its own facilities, be facilities located along its rights-of-way, its employees, and the public are worthy of protection.

Municipal organizations object to participating for one of two reasons: first, that the cost of membership is something they do not wish to bear; and second, that if they become members and thus receive notice of intent to excavate, they will have to have someone respond to the call and locate their facilities.

As to the former objection, the cost of membership is very small to most municipalities, normally consisting of the installation of a PC or a fax machine (which most of them have already), or something as simple as answering a telephone during normal business hours, and some one-call systems subsidize membership for small municipalities. Those municipalities with no underground facilities don't need to be members.

As to the latter, it is true that if you receive a call you have to do something about it. But the call, for example, can be tied into the excavation permit process, which most municipalities have, and this can be an excellent way of increasing permit revenue to a degree which will more than offset the cost of participation.

For instance, when the City of Pittsburgh, Pennsylvania joined the PA One Fall System (POCS) its street opening permit revenue was about $7,000 annually (1972). Currently, as a result of tying the permit application and one call notification together, the city generates in excess of $350,000 annually in revenue at a cost of about $4,300 for POCS membership—the cost based on nothing more than the number of calls it receives from excavators.

There is no compelling argument against participation in a one call system, and that there are, at present, creative, workable solutions which can make participation not only painless but even profitable to governmental facility owners and operators.

Other groups, such as railroads and on-shore natural gas producers, periodically have expressed concerns regarding "routine maintenance" and gathering lines, respectively, asking that they not be required to call or participate in one-call systems, as the case may be. We believe that any blanket exclusion is a bad idea, especially if the facility which could be injured by excavation (in the absence of notice) has the potential of causing any significant damage to person or property. Permitting the States to consider whether and how to grant limited exemptions, for good cause shown, and in very restricted circumstances may be appropriate, but no more than that. Also, let me point out, their employees deserve the protection too.

NOTIFICATION

Those who intend to excavate, as that term is broadly understood, should give advance notice of their intent to do so, except in an emergency where notice is not practicable. Even in an emergency, however, notice should be given as soon as possible. The notice should be given sufficiently in advance so as to permit affected parties to come to the job site to locate their facilities prior to excavation. Normally, this means that the notice should be given two or three business days in advance. The notification process ought to be easy to use for a project designer or excavator, and this generally should be through the use of a toll-free or "800" number directly to the one-call system serving the area where excavation is to occur. (We do not know whether the technology now exists to permit a single national one-call telephone number. This is a useful goal, however and it should be pursued in stages. Certainly there ought be a national information number to permit excavators to ascertain the correct "800" number for the state in which they intend to excavate, and each state should have such a number, which should be reachable from any other state. AT&T already maintains "800" directory assistance with standardized state listings; therefore, there would be no cost involved.)

Further, the excavator, when calling, must give specific information as to the job site and the nature of the work to be done. Finally, the excavator then must begin work, allowing the location of any underground facilities, in a timely fashion so that the any location marking does not become erased, stale or outdated due to the passage of time. The time-frame in the bill is appropriate.

EDUCATION

No one-call system will work unless it is used. Moreover, it misses the mark to consider one-call as anything other than an important part of an overall damage prevention program. The keys are to get excavators and facility owners/operators to become aware of their obligations, and to get them to see that it is in their own interest, as well as that of their employees, and the public, that they participate.

One-call systems and those who support them have comprehensive public education programs, which have been underway for years. Meetings with contractor groups, public service announcements, advertising media, presentations at trade shows, public proclamations, and virtually anything else which can be envisioned have been used. Some systems have regional representatives who go from place to place meeting with groups of excavators and facility owners/operators, and making presentations throughout the year.

OCSI has had an annual symposium on the subject since 1975, and this year's symposium, held in April in Minneapolis, attracted some 650 attendees over a five day period. Participants included one-call system operators, facility owners and operators, contractors, insurers, and federal, state and local officials.

The Federal agency that is given jurisdiction in this area should have a role in he public education process. We have found over the years that if people are educated they generally will participate, because intuitively the whole process makes sense. It makes sense from the standpoint of safety; it makes sense from the standpoint of saving time in the long run (quite apart from safety, if you don't call, begin construction, and find an underground facility, everything comes to a halt until you figure out what to do about it); and it makes sense from an economic standpoint (construction delays cost money and money, as does litigation).

Those who do not call find insurers lose patience and cancel's one's insurance. Without insurance, contractors can't bid on public contracts, and many private parties won't hire them either. It's a quick, painful way to go out of business. Not surprisingly, the insurance industry is a supporter of one-call damage prevention programs. So are responsible contractor organizations, as it's a good way of getting rid of competition from non-responsible excavators, as well as the bad press from the damage caused by those who don't call.

ENFORTERNENT MECHANISMS

Education works in the overwhelming majority of cases, and it can be made to work better with some constructive assistance from the Federal authorities. Nevertheless, there is a need for sanctions in instances where parties choose to ignore their obligations.

We strongly believe that sanctions should never become the tail that wags the dog; that the most effective sanctions are promptly imposed administrative sanctions, injunctive relief, and tort recovery limitation; and that civil and criminal penalties should be reserved for continuing, egregious violations of key provisions of the law.

Most States have some form of civil penalty; some have criminal penalties as well. A few (Massachusetts, Connecticut, and New Hampshire) have systems of administrative sanctions or citations.

Civil and criminal penalties, by and large, are not used, generally because much remedies are cumbersome, difficult to invoke, and require the cooperation of the State criminal justice system. Most State attorney general and district attorney offices are not interested in pursuing one-call violations unless there has been loss of life, significant personal injury, or significant property damage; they simply have too many other things to do. By the same token, they do not wish to delegate their enforcement powers to one-call systems, or facility owners and operators.

Moreover, we, as one-call systems, do not want to become enforcement agencies; it would confuse the public as to our role, distract us from what we do, and lead to fewer calls. We do see our role as providing information to parties to permit the law to be enforced by those with the authority to do so. Currently, we provide information in matters of civil litigation, and in the periodic requests for injunctive relief which are filed by facility owners and operators against excavators. We have done so, and can continue to do so in administrative, civil and criminal cases as well.

While there certainly have been and will be cases where the full force of the law must be employed, a well-designed and properly-run system of citations or other timely administrative sanctions remains the most likely to produce the results sought. The penalties do not need to be large, perhaps in the range of $100-$500 (with $250 a good mid-point), analogous to a traffic ticket.

Most people do call, and most people do respond and locate. The ones who do not; either are exempted by law from so doing, don't know of their obligation to call, or have a reason for not calling. Withdrawing exemptions will take care of the first group. Education generally will take care of the second group. Citations, perhaps after a warning, will get the attention of most who don't want to call, and may outweigh the reason for not calling—particularly if they are progressively increased. Moreover, a State could choose to combine citations with its bidding process on State contracts, and suspend or limit the right to bid in the event of a pattern of not calling. Insurance carriers could be required to be notified, so that they might apply additional pressure by threatening cancellation.

Restricting tort law recovery by persons whose failure to notify or to participate has caused damage, is another very effective means of compelling compliance. This system exists in Pennsylvania, for example, and has been a useful means of getting the attention of people in a creative way. It does not depend upon reference to the criminal justice system, and can be used by any excavator or facility owner in an appropriate case. It holds out both a benefit—a positive—to encourage participation and a detriment—a negative—to discourage lack of participation.

One important consideration is the Law of Unintended Consequences: any penalty system which is perceived as too severe may encourage persons who have not called to cover up the damage they've caused. That is perhaps the single most dangerous of all situations involving underground facilities. Any Federal or State scheme must recognize and address this fact by encouraging reporting, and by penalizing the failure to do so.

The bill should generally either require or encourage the States to create systems which provide for a mix of administrative, civil, and criminal penalties, plus injunctive relief upon application of an aggrieved party.

As we mentioned at the start, this bill, with some modest modifications, has the potential to be of real assistance to improving the process of damage prevention. We commend Sen. Bradley and his colleagues for taking a leadership role in Congress in this endeavor, and in joining with those of us who have dedicated a substantial portion of our professional lives to making one-call work for the public.

SPECIFIC COMMENTS REGARDING THE JULY 14, 1994 DRAFT OF HR 4394

As requested by the committee staff, following our comments regarding special provisions of the July 14, 1994, draft of HR 4394. It is our understanding that this draft is the document with which the committee will be working at today's hearing. We recognize that the thrust of the bill has changed since its original introduction, a fact which does not trouble us as long as the basic goals of comprehensive underground facility protection are met. With that in mind, and acknowledging that the July 14, 1994, draft does this, our comments are in the nature of clarification and tightening of certain ambiguities inherent in some of the words used.

1. Page 3, lines 5 through 9. This should include the persons leasing the property, but it is critical that the person be the operator. An operator—separate from the owner—doing excavation (or contracting for it to be done) is an invitation to confusion and disaster. Also, it needs to be clarified that there is no other underground facility located on the property than the one in question.

2. Page 7, lines 4 and 5. We remain skeptical regarding the need for a provision permitting citizen suits. Although this is not a critical issue, in the absence of a state statute there will be no provision for citizen suits; conversely, if the state has passed a proper statute, there is no need for citizen suits.

3. Page 7, line 10. We agree that if the only facilities at the site of an excavation are those of the owner, and the owner is excavating near those facilities, a different situation is created. The problem, however, is ascertaining that there are, in fact, no other underground facilities in the area of excavation. This results in the proverbial "circular argument." How is an excavator going to know that there are no other underground facilities at the site of excavation unless he calls? We do not think it to be a very good idea for the excavator to simply take the word of the party with whom it has contracted to do the excavation. There are enough examples of persons stating "there's nothing around here for you to worry about!" just before somebody hits a gas line with a backhoe, to make this a continuing bad idea. Advice like that is why one-call came into existence. We likewise are troubled by the words "at the site" as they may create too narrow an area of concern. We believe it would be better to say "at and near the site."

4. The penalty provisions contained at lines 13 through 24, page 7, and lines 1 through 12, page 8, are more appropriately stated than previously was the case, and satisfy the concerns we expressed at the House Energy and Power Subcommittee hearing. Our only question is whether the state's authority to impose criminal sanctions is thereby limited to the instances set forth in lines 20 through 24. page 7. We think this is a good limiter, but the states may choose to adopt lower threshold limits.

5. Having DOT come up with a modified "program" is a good idea and addresses many of the concerns raised by other parties throughout these proceedings. We remain in favor of statutory language which does not attempt to specify precise actions in every instance.

One Call Legislation

Mandatory One Call Participation

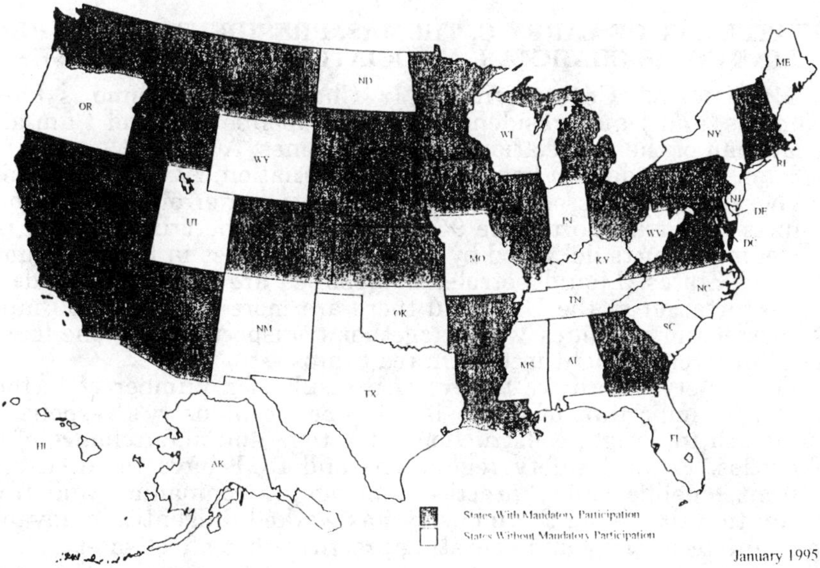

One Call Centers by Type

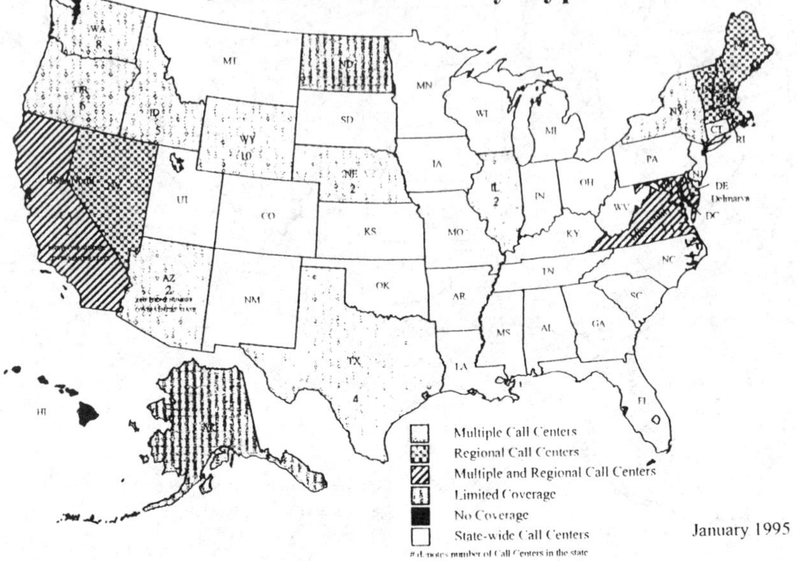

[Miscellaneous information may be found in the committee files.]

Senator EXON. Thank you very much, Mr. Boswell. I appreciate that. Mr. Thomas.

STATEMENT OF LARRY C. THOMAS, PRESIDENT, BP OIL PIPELINE CO. AS CHAIRMAN, ASSOCIATION OF OIL PIPE LINES

Mr. THOMAS. Good morning, Mr. Chairman. My name is Larry Thomas, and I am president of BP Oil Pipeline Co., and I am also chairman of the Association of Oil Pipe Lines, AOPL.

I am here today on behalf of that association. AOPL is an unincorporated trade association of 78 common carrier oil pipeline companies that transport some 95 percent of all the crude oil and petroleum products delivered by interstate pipelines in the contiguous United States. Liquid petroleum pipelines are the safest mode of transportation in the U.S., and there are more than 220,000 miles of petroleum pipelines which together transport almost one-fourth of all intercity freight in the United States.

Our safety record we believe is a result of a number of factors, the most important of which is the pipeline industry's responsible approach to design, construction, operation, and maintenance of its facilities. Federal safety regulations and DOT have required consistent pipeline safety practices among the companies, and from State to State. Also, the industry has worked diligently to advance pipeline technology and operating practices that affect safety.

In the last 40 years, we have seen great improvements in corrosion control, quality of pipe and equipment used in pipelines, integrity testing methods, and in computer instrumentation for pressure control and pipeline monitoring, and that includes leak detection, yet we still have a major cause of major pipeline accidents, and that has been third-party excavator damage, an activity over which the pipeline operator has very limited control.

Despite regular aerial and, in some cases, surface inspection—walking pipeline rights-of-way—aboveground signs marking the pipeline routes, active public education programs regarding the hazards of digging around pipelines, and existing one-call systems, third-party damage continues to be a major problem.

One-call systems have become a proven method of alerting the excavator to the existence and location of underground facilities, and it has been used effectively. It is, and has been, our position that one-call systems should be encouraged and supported.

Today, my remarks will be directed to the amendment in the nature of a substitute to H.R. 4394, which I understand is to be offered by Congressman Phil Sharp. As I have stated, the oil pipeline industry strongly supports one-call concepts and systems.

Some concerns have been expressed regarding inclusion of gathering and flow lines not regulated by the Department of Transportation under the Hazardous Liquid Pipeline Safety Act. A meeting is scheduled, in fact, for this afternoon at the API to attempt to reach an industry consensus on this matter. I will be happy to provide this committee with a report on the results of that meeting.

Aside from these concerns, we support broad-based participation in one-call systems by both owners of underground facilities, including all hazardous liquid pipelines regulated by the Department of Transportation and excavators. We are especially pleased that under this legislation, Government entities, and public utilities, as well as private underground facility owners, are required to participate in the proposed one-call systems.

We believe that a single 1-800-type number would facilitate communication and public education, including public service announcements, and notification along the right-of-way, thereby improving the use and effectiveness of one-call systems.

We note that the bill calls for the creation of State one-call programs either through the establishment of a new program or through modification or improvement of an existing program, and that a State program may be implemented by a nongovernmental organization. This grandfathering of existing programs we believe is a very positive feature.

It is also advisable for the DOT to craft a model program, using its expertise and the resources of others involved in one-call, to assist the States in their efforts.

Mr. Chairman, this concludes my testimony. It is a pleasure to appear before you and your committee to discuss the important subject of pipeline safety. I assure you, as the Association of Pipelines, we will work with you.

[The prepared statement of Mr. Thomas follows:]

Prepared Statement of Larry Thomas

My name is Larry Thomas and I am the President of BP Oil Pipeline Company, and Chairman of the Association of Oil Pipe Lines (AOPL). I am here today on behalf of that Association.

AOPL is an unincorporated trade association of 78 common carrier oil pipeline companies that transport some 95 percent of the crude oil and petroleum products delivered by interstate pipelines in the contiguous United States.

Liquid petroleum pipelines are the safest mode of transportation in the United States. There are more than 220,000 miles of petroleum pipelines, which together transport almost one-fourth of all intercity freight in the U.S.

Our safety record is the result of a number of factors, the most important of which is the pipeline industry's responsible approach to the design, construction, operation and maintenance of its facilities. Federal safety regulations require consistent safety practices among the pipeline companies and from State to State. Also, the industry has worked diligently to advance pipeline technology and operating practices that affect safety. The last forty years have seen great improvements in corrosion control, quality of pipe and equipment used in pipelines, integrity testing methods, and in computer instrumentation for pressure control and pipeline monitoring, including leak detection.

There are many very real incentives for this industry to operate safely. These include responsibility to the public for potential loss of life, injury, property and environmental damage; significant capital investment; and the need to attract investor, employee and general public good will. Add to this one obvious fact—a disabled pipeline does not transport oil or refined products.

A major cause of pipeline accidents has been third-party—excavator—damage, an activity over which the pipeline operator has limited control. In order to minimize such third-party accidents, present DOT regulations provide for routing proposed pipelines around congested residential and industrial areas, wherever possible. However, many pipelines were constructed years ago in rural areas, which have since become populated. Despite regular aerial and, in some cases, surface inspection of pipeline rights-of-way, above ground signs marking pipeline routes, active public education programs regarding the hazards of digging around pipelines, and existing one-call systems; third-party damage continues to be a major problem.

ONE-CALL SYSTEMS

One-call systems have become a proven method of alerting the excavator to the existence and location of underground facilities and have been used effectively. It is, and has been, our position that one-call systems should be encouraged and supported.

THE COMPREHENSIVE ONE-CALL NOTIFICATION ACT OF 1994, H.R. 4394

Today, my remarks will be directed to the Amendment in the Nature of a Substitute to HR 4394 which I understand is to be offered by Congressman Phil Sharp.

As I have stated, the oil pipeline industry strongly supports one-call concepts and systems. Some concerns have been expressed regarding the inclusion of gathering and flow lines not regulated by the Department of Transportation under the Hazardous Liquid Pipeline Safety Act. A meeting has been scheduled for this afternoon at the American Petroleum Institute to attempt to reach an industry consensus on this matter. I will be happy to provide the Subcommittee with a report on the results of that meeting.

Aside from these concerns, we support broad-based participation in one-call systems by both owners of underground facilities, including hazardous liquid pipelines regulated by the Department of Transportation (DOT), and excavators. We are especially pleased that under this legislation government entities and public utilities, as well as private underground facilities owners, are required to participate in one-call systems.

As to the benefits of a nationwide toll-free telephone number, we understand that what is envisioned is a single number in use nationwide that would access the local one-call system—much as dialing 411 accesses the local directory assistance. We believe that a single 1-800 type number would facilitate communication and public education, including public service announcements and notification along the right-of-way, thereby improving the use and effectiveness of one-call systems.

We note that the bill calls for the creation of State One Call programs either through the establishment of a new program or through modification or improvement of an existing program, and that a State program may be implemented by a non-governmental organization. This "grandfathering" of existing programs is a positive feature.

We also agree that the criminal penalties and enforcement section, which leaves to the States the discretion to prescribe the actual penalties for violations, is appropriate.

It is also advisable for the DOT to craft a model program utilizing its expertise, and the resources of others involved in one-call to assist the States in their efforts.

With respect to State grants, we understand the funds contemplated for grants to States to establish one-call systems are currently in escrow. Further appropriation of funds should not occur until the next session of Congress when the Pipeline Safety Act Reauthorization is again considered.

Mr. Chairman, this concludes my testimony. It is a pleasure to appear before your Subcommittee to discuss the important subject of pipeline safety. I assure you of our commitment to work closely with you and your staff as the legislative process moves forward. I would be pleased to answer any questions you or your colleagues may have.

Senator EXON. Mr. Thomas, thank you very much. Mr. Devaney.

STATEMENT OF PAUL E. DEVANEY, MEMBER, TECHNICAL STAFF, BELLCORE, AS CHAIRMAN, UTILITY LOCATION & COORDINATION COUNCIL, AMERICAN PUBLIC WORKS ASSOCIATION

Mr. DEVANEY. Good morning, Mr. Chairman, and I certainly thank you for this opportunity to provide testimony regarding the Comprehensive One-Call Notification Act of 1994. As a member of the technical staff of Bellcore, I actively participate on a national level with associations, Federal agencies, and standards bodies involved with damage prevention, rights-of-way, and legislative and regulatory matters, as well as other industry-related activities.

I currently serve as the chairman of the Utility Location and Coordination Council of the American Public Works Association, and it is in this capacity that my testimony is being provided today.

I believe that you have correctly identified, in Senate bill 2101 and also in the House of Representatives bill 4394, that damage to infrastructure is not just a pipeline problem, it is a persistent, nationwide problem affecting all utilities with below-ground facilities. The Utility Location and Coordination Council supports the efforts of the comprehensive one-call notification legislation.

Approximately 20 million miles of pipes, cables, and wires make up this country's utility infrastructure. These facilities, found both above and below the ground, help to provide for the excellent standard of living that can be found in most areas of the United States.

Like many things, people have a tendency to take this standard for granted, and rarely are aware of the sometimes invisible support systems that make our standard of living possible. We all rely on the infrastructure to support and improve the quality of our lives, and to underpin our economy.

Today, construction and maintenance excavations have made damage to facilities a much-too-familiar occurrence. The most severe, dramatic damage will draw attention, while the large majority will just be repaired and replaced. Additionally, some damage is not always immediately evident, and the full impact of this damage can extend far beyond a sudden facility failure.

Damage to a facility can affect worker safety and pose potential safety problems to other operations such as aviation, vehicular traffic, and emergency services. It can also result in time-consuming and costly disruptions to the public and business community, as well as cause long-term damage to the underground facilities.

The cost of this damage and disruption is ultimately borne by the taxpayer/ratepayer, and many times these payers are from a different geographic area that is separate from the location of the actual destruction.

There is an ever-increasing need for protection of our infrastructure for critical services. Future systems such as monitoring or transmission of medical information over communications facilities, and technology advances such as the intelligent highway vehicle system, which will require its own infrastructure of sensors and computer links, are becoming a reality.

The infrastructure of this Nation, by and large, has been built and owned through subsidy. Whether the subsidy was through the sale of stocks or bonds, or generated by rates or taxes, it is of little concern. The universal protection through the best-known means is what is at issue as we address the need for the important Federal legislation.

Mandatory participation without exception in one-call systems is needed today. The American Public Works Association acknowledges that one-call notification systems provide a communication as well as a coordination link between excavators and utility owners. This link has proven to prevent damage to utilities, reduce potential hazards to workers, and reduce service disruptions.

National standards are needed. The key is also to have uniform criteria and standard elements established and maintained through the combined efforts of the facility operators, excavators, one-call notification operators, and State and local governments. This forum could also provide a method to change and update these criteria without the need to constantly modify damage prevention laws.

The use of an existing standard, such as the APWA Uniform Color Code, as well as other industry and national standards, would help to maintain this uniformity. It would make little sense to have a telephone number that would help a contractor working in multiple States if they needed to carry around varying State statutes and standards books. There are many examples of existing standards in use today that use this format and provide guidance to manufacturers, consumers, and the general public.

Education concerning damage prevention for excavators, whether they be contractors, homeowners, utilities, public agencies or officials, should be available and up to date if changes in current behaviors and attitudes are to be made. Making a notification call is only the first step in the damage prevention process. Recognizing and understanding the proper techniques and responsible actions required for damage prevention is the key.

The use of a nationwide toll-free telephone number could help to reduce advertising costs as well as promotion materials, and would be a convenience to contractors working in multiple States. The availability of this type of service should be investigated, and consideration should be given for finding an easy-to-remember number.

For penalty systems, our recommendation would be for a uniform system of citations or administrative sanctions. In our view, these would likely produce the best compliance among equipment operators and excavators.

We are not suggesting that there be no stringent penalties, but that a certain balance of consequences must be available, with severe penalties used as a last resort for those who either willfully or continually damage facilities. Any penalty system that is perceived as too severe may encourage violators to hide or cover up the damage they have done, and thus create the potential for a catastrophic event.

Mr. Chairman, the Utility Location and Coordination Council of the American Public Works Association supports comprehensive one-call notification legislation for the protection of underground facilities. I believe the July 14, 1994 discussion draft of H.R. 4394 takes great strides in addressing the many complex issues associated with this legislation. Adoption of the recommendations contained in my testimony would, we believe, further improve the bill.

I appreciate this opportunity to address the committee, and look forward to working with you on this legislation as it moves forward.

Senator EXON. Mr. Devaney, thank you very much. We have a rollcall vote on, so I will have to move forward rapidly and not ask as many questions as I would have. I simply would tell all of you on this panel and the preceding panels that we very likely will have additional questions for the record which we will send you, and ask for you to answer as quickly as you can. This will be for the official part of the record, because we want all of your suggestions and a full record on this.

There are two brief questions which I think are somewhat important. I ask one or two of you to comment briefly, if you could. Is a nationwide toll-free number absolutely necessary for proper functioning of a one-call system? Is it possible the system could function just as well with regional or Statewide telephone numbers? I would ask you to comment on that, Mr. Devaney, because you have brought it up.

Mr. DEVANEY. I would like to defer that question to Bill Kiger, who has been running a successful one-call system for many, many years, and I think that he would seize that opportunity.

Mr. KIGER. Mr. Chairman, I think that that is not a necessary component of an effective system. The States that have nationwide availability of their number are adequate to do that.

There are States that need to upgrade their 800 number to that level of service, but the existing States, only one does not have an 800 number that they could use, and they have spent thousands and thousands of dollars, if not hundreds of thousands of dollars, over the past 25 years promoting those numbers.

Senator EXON. Do the rest of you agree with that?

Ms. DENSLOW. Yes.

Mr. THOMAS. Yes.

Mr. MCNIEL. Yes.

Senator EXON. My next question: please provide the committee with your recommendations briefly as to how States could equitably set fees necessary to operate the one-call system. Is there any magic formula in that regard?

Mr. BOSWELL. Mr. Chairman, generally the fees are set by the systems themselves. What the States have done traditionally over the years is establish that there shall be one-call systems. Most of the systems are not-for-profit corporations, and they set the fees themselves for their members, so the members vote as a group on what the fee structure will be, and I see no reason at all why that should not continue into the future. It has worked very well so far.

It is a democratic way of approaching things, and it is really the best way of proceeding.

Senator EXON. Is there general agreement on that?

Ms. DENSLOW. Yes.

Mr. THOMAS. Yes.

Mr. MCNIEL. Yes.

Senator EXON. I wish I could get my colleagues in the Senate to agree as much as you people seem to agree. It would sure make life a lot easier.

With that, I thank all of you for being here. It has been very, very helpful to us. I assure you we are going to move forward on this as rapidly as we can.

Thank you, and with that, we are adjourned.

[Whereupon, at 10:55 a.m., the committee adjourned.]

APPENDIX

Prepared Statement of Senator Pressler

Mr. Chairman, I thank you for holding this hearing on the Comprehensive One-Call Notification Act of 1994. It is unfortunate that it took a tragedy to have Congress move forward on this legislation.

What happened in New Jersey could have happened anywhere in this country. There are millions of miles of underground pipelines, cables, and wires that are essential to the infrastructure of this country. This infrastructure moves energy resources, information, and other services that have become our lifestyle norm.

In 1992, during the reauthorization of the Pipeline Safety Act, I worked to improve the safety of the hazardous liquid pipelines. My legislation provided for additional safety inspectors in order to reduce the potential for leaks. I was pleased this committee worked with me to improve the safety features at that time.

At that time, it was brought to my attention that my home State of South Dakota did not have a one-call system in place. Therefore, I am pleased to be able to report to my colleagues today that South Dakota is working on a one-call system that will be operational by January 1, 1995. The rulemaking process has just been completed in conjunction with the Public Utilities Commission. Mandatory participation by contractors, utilities, municipalities, and others in the one-call system will reduce the potential for any tragedy to occur.

Any person digging underground will be required to call 48 hours prior to digging. South Dakota will have just one center to deal with calls.

Thank you, Mr. Chairman.

Prepared Statement of Robert S. Cave, Executive Director, American Public Gas Association

The American Public Gas Association ("APGA")[1] appreciates the opportunity to submit written comments on a national, comprehensive one call notification system and commends the Subcommittee on Surface Transportation for moving quickly on this important public policy issue. APGA supports Congressional efforts to enact new legislation requiring notification to and participation with one-call systems and looks forward to working with Congress and the Executive Branch in implementing this legislation.

APGA requests Congress to develop expeditiously a one-call program which is thorough, well-understood, and widely publicized. APGA is committed to supporting a one-call system and has been working with the Department of Transportation's Office of Pipeline Safety over several years to develop guidelines for a one-call system.

The time has come for Congress to enact legislation that ensures all states have laws which:

- establish a one-call system compelling all excavators, including state and local governmental entities, to call before they dig;
- require all underground facilities used for gas, electricity, communication, petroleum, water and sewage to participate so excavators can secure all of the necessary information about a site with one telephone call;
- use a telephone number (such as 344–"DIG") which functions similar to the 911 emergency number. Establishing a single telephone number will make the one-call system easy to use.
- promote public awareness of the benefits of "calling before you dig."

[1] APGA is the service organization for this nation's community-owned, not-for-profit public gas utilities that provide retail natural gas service to millions of consumers, businesses, and industries in 34 states. Public gas systems serve approximately 3.8 million of the nation's gas consumers.

• Also, federal legislation should leave it to the states to develop their own sanctions and penalties for those not following the law. Several options for sanctions and penalties to be considered include: civil penalties; administrative fines; judicial injunctive relief; contempt charges (for failure to comply with a judicial or administrative order to comply); strict civil liability; civil liability "multipliers"; excavator disbarments; and publication of lists of violators. Regardless of the method used, each state should have some penalty for not calling or notifying a one-call system.

To maximize the effectiveness and consistency of state and local one-call programs, there should be no exemptions to these standards. In addition, states, local utilities, and excavators should be encouraged to participate by developing a one-call notification program that both works best for them and meets the national standards.

To influence many states, counties, municipalities and excavators to participate in a one-call program, APGA suggests that Congress consider linking federal funding for state and locally sponsored highway, railroad, aviation, and transit projects to participation in a one-call program. Thus, a state, county or municipality would be required to participate in a one-call program in order to receive federal funding for its capital improvement project. Similarly, all excavators could be required to have called a one-call program prior to construction in order to receive payment for work performed under a federal program. APGA encourages active federal oversight to insure that states, counties, municipalities, and excavators are participating in a one-call program.

APGA again thanks the Subcommittee on Surface Transportation for holding a hearing on this important issue and urges the Subcommittee to incorporate our comments for one-call notification into legislation being considered. On behalf of public gas systems across the country, APGA looks forward to working with the Subcommittee to develop and implement a safe, effective one-call notification program.

PREPARED STATEMENT OF DANIEL E. CRAWFORD, PRESIDENT, NETWORK SERVICES, MCI COMMUNICATIONS CORP.

I am Daniel E. Crawford, the President of Network Services of MCI Communications Corporation (MCI). In that capacity, I oversee all technical and operational aspects of MCI's network, including repair of severed or otherwise damaged transmission cable. I have previously served as the head of the Network Reliability Council's (NRC) Focus Team on Fiber Cable Systems and was the principal author of the article "Fiber Optic Cable Dig-Ups: Causes and Cures" which was published in the NRC's June 1993 report "Network Reliability: A Report to the Nation." A copy of that article is being submitted as an attachment to my testimony and I ask that it be made part of the hearing record.

On behalf of MCI, I wish to thank the Committee for affording me the opportunity to submit this statement regarding S. 2101, the "Comprehensive One-Call Notification Act of 1994." MCI commends you, Mr. Chairman, Senator Bradley and the entire Committee for your efforts to fashion responsible legislation to address the concerns stemming from dig-ups of underground facilities, including communications facilities.

MCI celebrated its 25th anniversary in 1993. During those years we converted a network which began with a single microwave link between Chicago and St. Louis to one which now provides service to the nation over high capacity optic fiber. MCI's network now utilizes 21,000 miles of optic fiber cable nationwide, and approximately 97 percent of that fiber is in underground facilities. Maintaining the integrity of those underground facilities is essential to the continued reliability of our network which in turn is inexorably tied to the quality of service our customers receive. In the vigorously competitive long distance market, a competitor will not long survive if its customers cannot make calls due to a network outage resulting from a severed cable.

We view S. 2101 as a particularly important initiative because we believe that although our fiber network is underground and out of sight it should never be out of mind. The optic fiber located in these underground storage facilities, although composed of glass strands only slightly thicker than human hair, nonetheless comprises the backbone of the much-heralded information superhighway. Clearly, as our economy becomes increasingly information based, maintaining the integrity of the highways upon which that information flows takes on an added degree of importance in furthering our nation's economic progress. Outages related to cable failures cost carriers and customers alike and take time to repair. The industry average time for service restoration is 5.2 hours and the average time for cable repair is 14.2 hours.

More importantly, and just as with pipelines and other utilities which occupy underground facilities and rights of way, ensuring the integrity of underground communications facilities has broad-based public health and safety implications. Whether related to encrypted communications of the United States Armed Forces, telemedicine, nuclear power facilities or other vital societal functions (such as 911 database access), an interruption in service precipitated by a fiber cut can dramatically affect the lives of millions of Americans.

The Fiber Cable Focus Group of the NRC, which I headed, collected and analyzed data on service interruptions traced to cable failures caused by dig-ups. The Group also reviewed the practices of other utility companies and outside groups such as "one-call" associations. The first recommendation made by the Group was to strengthen utility damage prevention legislation and enforcement—so called "call-before-you-dig" or "one-call" legislation.

The priority given call-before-you-dig measures by the Fiber Cable Focus Group can easily be accounted for. Nearly 100 percent of network failures studied by the Group resulted from a broken or damaged cable. Dig-ups are the largest single cause of fiber cable failures and account for nearly 60 percent of the cable failures reported in the industry. When we examined the root causes of the dig-up occasioned fiber cuts, we noted that 33 percent resulted from the excavators' failure to notify the owner of the underground facility before digging started.

MCI's internal numbers indicate that properly administered call-before-you-dig systems are both effective and efficient in drastically reducing the number of outages resulting from cable cuts. In 1993, for example, MCI received approximately 569,365 notifications from call-before-you-dig or one-call systems. We sent engineers to the field to locate cable on 79,711 occasions. Of that number of excavations, only once was MCI's cable damaged. It is our experience that the critical element of these systems is the notification process. Since facility operators have both the expertise and incentive to protect their facilities, they should continue to be vested with a degree of discretion as to whether, given the site of the excavation, they need to send personnel to the field to locate and mark their facilities.

Federal legislation is needed to ensure that call-before-you-dig systems are uniformly developed and administered across the country. Although over 40 states in 1993 had damage prevention laws requiring notification prior to excavation, the Fiber Cable Focus Group found these laws were generally weak, not adequately enforced and provided little in the way of deterrence through punitive damages. That is why federal legislation is so necessary.

Let me demonstrate the pressing need to address the one-call issue on the federal level through some examples. A new one-call statute in Kentucky exempts all utilities and their contractors from one-call compliance. This amounts to approximately one-third of all excavations Statewide. In Georgia, North Carolina and South Carolina, the states' Department of Transportation are exempt.

We view statutes with these kinds of broad exemptions as creating a situation worse than having no statute at all. At least, in the absence of a statute, there is an industry standard to call before digging. But granting exemptions only promotes more reckless excavations since there is no recourse for anyone harmed as a result of a digup by an exempt contractor otherwise in violation of the state one-call statute.

MCI believes that all excavators should be equally subject to call-before-you-dig requirements. The Fiber Cuts Focus Group found that excavators working for local telephone companies, electric, gas, water and highway department contractors were all involved in significant degrees with cable cuts. Only approximately 4.5 percent of cable cuts resulted from dig-ups by private landowners.

S. 2101 has played an important role in alerting industry to the importance of one-call issues, and the bill strikes a number of important balances. It seeks standardization of one-call systems across the country, not by establishing another federal bureaucracy, but rather, through setting minimum standards for state one-call systems. Each state, however, remains free to implement its own system once approved by the Secretary of Transportation. S. 2101 imposes obligations on both the excavator and the facilities operators and in this regard, the bill establishes a framework for their cooperation.

However, in the absence of voluntary cooperation, S. 2101 also provides for proper penalties for repeated violation of the one-call provisions. This form of deterrence is vital to the success of any program designed to require excavators to call before they dig.

MCI is enthusiastic regarding the prospects for call-before-you-dig legislation and looks forward to working with you, Mr. Chairman, Senator Bradley and each of the Members of the Committee to further refine and make needed changes to S. 2101. Toward that end, we would be happy to supply the Committee with whatever additional information or resources it may require as it continues to consider this important legislation.

Thank you.

["Fiber Optic Cable Dig-Ups—Causes and Cures," by Dan Crawford, President, Network Services, MCI Telecommunications Corp., may be found in the committee's files.]

QUESTIONS ASKED BY SENATOR EXON AND ANSWERS THERETO BY MR. HALL

NATIONWIDE TELEPHONE NUMBER FOR ONE-CALL

Question. Is a nationwide, toll-free number necessary for the proper functioning of one-call systems? Is it possible that the system could function just as well with regional or statewide telephone numbers?

Answer. We do not believe a national one-call, toll-free number for notifying of planned excavations, is essential. However, such a telephone number would be of assistance to excavators for obtaining one-call center numbers when they intend to work in states other than those in which they reside and for pipeline operators that have facilities in more than one state.

GRANTS TO DEVELOP ONE-CALL SYSTEMS

Question. The bill provides for grants to States to develop or upgrade existing one-call systems. Should Congress consider making non-government entities running one-call systems eligible for such grants?

Answer. We do not believe it necessary to provide grants to organizations other than the States. Operators of buried facilities now pay the cost for one call notification systems through fees or assessments based generally on the demand they impose on the center.

PARTICIPATION IN ONE-CALL SYSTEMS

Question. Does the National Transportation Safety Board (NTSB) favor participation of all facility owners and operators in a one-call system? If not, under what circumstances should a facility owner or operator or an excavator be exempted from participation?

Answer. The Safety Board believes it essential that all excavators and operators of buried facilities, whether private or public organizations, participate in each state's excavation damage prevention program. This is the only way these systems will be able to achieve their full potential for improving public safety and for reducing the costs and inconveniences needlessly borne by the public.

The only exception we believe might be made without adversely affecting the safety objective of an excavation damage prevention program is where the only buried facilities are located on land controlled by the operator of the facilities and which only serve the operator. Such an exception would exclude large industrial complexes and gathering fields where the operator has control over all activities, including excavation. The Safety Board believes that an exclusion should be provided only where it can be demonstrated that the exclusion will not adversely affect public safety.

ADDITIONAL STEPS TO PREVENT PIPELINE DAMAGE

Question. In addition to this legislation, what other steps might Congress take to aid pipeline damage-prevention programs?

Answer. The Congress should establish deadlines for the Department of Transportation (DOT) regarding rulemaking actions mandated by this legislation. Expedited rulemaking action is needed to define the specific provisions that the States must incorporate into their programs before receiving any of the available grant monies. Any delay by the DOT in establishing its regulations will unnecessarily delay implementation by the States. Additionally, the Congress needs to ensure that adequate grant money is appropriated for the States to expeditiously make necessary program modifications.

The Safety Board is co-hosting with the DOT an excavation damage prevention workshop on September 8 and 9, 1994, which will bring together several hundred individuals experienced in the operations of excavation damage prevention programs. From that workshop, we will learn if there are other elements necessary for developing effective programs and we will gain more detailed information on each element. We believe that this workshop will provide for both the Office of Pipeline

Safety (OPS) and the Safety Board the information needed to motivate future improvements in these programs.

MODEL ONE-CALL PROGRAMS

Question. Should the bill's provisions for a model one-call system be changed to require the Secretary of Transportation to decide what a model one-call system should provide?

Answer. We believe that the bill must include basic requirements (as it now does) for all operators of buried facilities and all excavators. Beyond the basic provisions now included in the bill, the Secretary is given sufficient authority to define through the rulemaking process all other aspects that should be included in a model system.

Again, it is hoped that in the excavation damage prevention workshop on September 8 and 9, 1994, the Safety Board will develop elements necessary for effective programs. We believe that this workshop will provide for both the OPS and the Safety Board the information needed to motivate future improvements in these programs.

PIPELINES IN DENSELY POPULATED AREAS

Question. In the broader context of pipeline safety, NTSB has suggested additional safeguards for pipelines in densely populated areas. Please elaborate for the Committee as to what additional safeguard NTSB believes are necessary for pipelines in such areas.

Answer. The Safety Board has identified the need for additional safety standards for pipelines in densely populated areas. The Safety Board's investigation of the March 23, 1994, gas transmission pipeline rupture at Edison, New Jersey, and previous Board investigations addressed this issue. As you know, in the Edison accident, 23 residents were injured and taken to area hospitals and eight nearby multistoried apartment buildings were destroyed with six others receiving substantial damage. The Safety Board previously investigated an accident in Mounds View, Minnesota in 1986 that involved the rupture of a pipeline and spread of gasoline and fire in parts of six blocks that resulted in two fatalities and one injury. An accident in San Bernardino, California in 1989, also involved the release of gasoline from a pipeline and resulted in 2 fatalities, 19 injuries, and damage to 17 homes. Recommendations in both the 1986 and 1989 accidents called for the installation of remote-operated valves in highly populated areas. The recent Allentown accident highlights the potential danger of gas and vapors near tall buildings.

Liquid pipelines also traverse densely populated urban areas as did the natural gas pipeline at Edison. The close proximity of people to high-pressure liquid pipelines also creates needs for similar or better safety improvements. The need for routing, establishing rights-of-way widths, and providing educational information to assist local governments is even greater for liquid pipelines used to transport highly volatile liquids. As an example as to how far heavier than air gases might spread and the potential area of impact, propane gas escaping from a pipeline at North Blenheim, New York, flowed more than ½-mile from the rupture before it ignited resulting in two deaths, seven injuries, and more than $3 million in damages. Also, liquefied petroleum gas released from a pipeline near Whitharral, Texas, engulfed a ¾-mile by ¼-mile wide area before it ignited and resulted in five deaths and two injuries. In more densely populated area, hundreds of fatalities could result from similar accidents.

The Safety Board previously addressed the need to educate localities and the public on land-use planning of properties in the vicinity of such pipelines, especially for those pipelines used to transport highly volatile (HVL) liquids, materials such as LPGs and anhydrous ammonia. In a 1983 report on a liquid pipeline rupture that killed six persons and injured four others in a new developing community, the Safety Board concluded that the development "was planned by its developer and approved by local officials without consideration of the effect of the development on the * * * pipeline or the safety of the residents." As a result of the 1983 accident, the Safety Board recommended that several organizations notify their members of the accident and "urge them to cooperate with local government land planning and zoning agencies in the development and implementation of restrictions against the development of residential lots over pipelines transporting hazardous liquids or gases or of lots on which construction will necessarily encroach on easements for the pipelines."

In a 1989 Board accident report, the Safety Board concluded that the "City of San Bernardino's plan for land use did not address the hazards posed by the proximity of mainline railroads and of high pressure pipelines." A recommendation to the City asked them to revise the existing plan for land use to account for the location of high pressure pipelines. The National Association of Counties and the National League of Cities were requested to inform their members "of the land-use guidance for enhancing public safety contained in the National Research Council's Special Report 219, 'Pipeline and Public Safety,' and encourage them to develop and implement policies to protect public safety for lands adjacent to pipelines and railroads." The National Association of Counties published an article in "County News" on the report. The Safety Board received no response from the National League of Cities.

The Edison accident again shows the need to address public safety issues relative to lands adjacent to pipelines:

Natural Gas Pipeline Routing: When certifying new natural gas pipelines, the Federal Energy Regulatory Commission (FERC) requires pipelines to meet DOT's safety standards and therefore has a shared responsibility with DOT for safety. Decisions made by the Federal Energy Regulatory Commission (FERC) about changes needed for selecting pipeline routes and for determining the appropriate rights-of-way widths should include the potential hazards to public safety. The hazard assessments should be made not only of the blast damage effects, but also on the distance radiant heat may threaten buildings and occupants should a pipe rupture fully and the escaping natural gas be ignited.

Liquid Pipeline Routing: DOT could establish rights-of-way widths, and develop educational information to aid local jurisdictions. We believe that the DOT Offices of Pipeline Safety and Hazardous Materials (both a part of the Research and Special Programs Administration), could follow review procedures similar to those used by FERC for natural gas pipelines to develop routing criteria and minimum rights-of-way widths for each of the liquid products transported by pipeline. Pipeline planning zones that account for population density and product risk should be established based on the type of product, size of pipe, pressures, and topography.

Frequency of Internal Inspection: Internal inspection for both natural gas and liquid pipelines should be based on the risks posed by the pipeline to adjacent populations and environmentally sensitive areas. Factors that should be considered in assessing the risks should include the incident, operating, and maintenance histories of the pipeline segment; the properties of and the hazards posed by the products being transported; and the maximum operating pressure.

Public Education and Information: People residing within pipeline planning zones need periodic reminders informing them on how to recognize hazardous conditions and the actions they should take for their safety. Since 1984, and the investigation of the Odessa, Texas accident, the Safety Board has encouraged that information should be available to prospective residents and tenants before they commit to the purchase, lease, or rental of property within pipeline planning zones. This safety information needs to go to all tenants even though many rent houses or apartments for periods less than a year.

Hazardous liquid lines transport a variety of products that pose significantly different threats to public safety and to the environment. Pipeline planning zones should be established based on risks posed to public safety and the environment for each product transported by pipeline. The potential threat of a liquid released from a pipeline should be determined based on factors such as the properties of the most hazardous product transported in a pipeline, on the operating characteristics of the pipeline including the capability of the operator to promptly detect and stop the flow of product to the release point, on the topography along the pipeline, and on the existence of underground systems in which released materials may enter and flow long distances from the release site.

Automatic and Remote Operated Valves: Since 1971, the Safety Board has advocated the rapid shutdown of gas and liquid pipelines. Additional recommendations were issued in 1972, 1974 and 1979. In the Mounds View accident in 1986, remote operated valves were recommended based on the population at risk. In the San Bernardino accident in 1989, the 1987 recommendation was reiterated. In San Bernardino, another recommendation to RSPA called for the evaluation of check valves in combination with remote-operated valves for use in automatically controlling the release of products from pipelines.

In the Edison accident, the force the employees needed to exert to operate the valve wheel became increasingly greater as the valve neared closure. This was due to the increasing pressure difference between the pressures upstream and downstream of the valve as the downstream gas pressure vented from the rupture which was just downstream of the valve. As the valve neared closure, the pressure differential dramatically increased and made it impossible for the crew to close the valve. This problem was overcome by having workers close the next upstream valve. This equalized the pressures upstream and downstream of the nearest valve to the rupture and workers then were able to easily close that valve. Had the second valve upstream of the rupture been closed first, the flow of gas could have been stopped much earlier.

There are a number of alternatives to solve problems including back-up power valves, check valves and remotely operated valves. Check valves on one-way flow lines, in fact, are automatic closing valves. Used in combination with remote-operated valves, often can provide the best means for rapidly isolating a failed section of pipeline and may be a potential cost-effective alternative. Legislative provisions should not exclude the use of automatic valves, including check valves. In either instance, the valves installed should not reduce the ability to perform internal inspections of the pipeline.

Excavation Damage Prevention: Operators of pipelines in densely populated areas should have an active excavation-damage prevention program including full participation in one-call notification centers where available, periodic patrolling of right-of-ways to detect unreported excavation activities, and aggressive education programs to inform contractors and the public of the need to notify of planned excavations, how to protect the pipeline from excavation damage, and how to report immediately any contact with the pipeline by excavation equipment. Federal regulations now require natural gas pipeline operators to have excavation damage prevention programs and to participate in available one-call notification systems; however, this requirement is yet to be extended to liquid pipeline operators.

PENALTIES UNDER THE BILL

Question. In the past, NTSB has favored monetary penalties and injunctive relief against violators of one-call requirements. Are the penalty provisions in this bill in line with NTSB's suggestions?

Answer. The Safety Board has not addressed specific penalty provisions in its accident reports or safety recommendations.

The Safety Board has called only for effective enforcement actions. Compliance with laws is achieved most effectively when there is good reason to believe that to do otherwise will not be in one's best interest.

ONE-CALL VS. EXCAVATION PREVENTION

Question. Please elaborate on the differences in one-call notification systems and excavation prevention programs. Also, please detail for the Committee, the current role each plays in ensuring pipeline safety.

Answer. Excavation damage prevention systems (programs) include all elements necessary for limiting the potential of excavation damages to the infrastructure, as follows:

- State legislation that establishes the program purpose, who must participate, the operating performance parameters (including provisions for notifying of planned excavations or blastings), how the program is to be administered, the penalties for violating the program parameters, and the means of enforcement.
- System for administering the program. The program might be administered by a state agency authorized to perform surveillance of excavation and blasting activities, to issue citations on probable violation, and to make final determinations. Also, it might be administered through the courts by action initiated by an excavator or by a buried facility operator.
- Excavation notification center. A system, whether or not required by statute, intended to ease the notification burden for planned excavations. Such a center makes it possible for a single notice to the center to alert all participating facility operators. The ultimate for this system is to allow one notice to alert all buried facility operators and thus become a true "one-call notification system."
- Education/motivation actions to encourage pre-notification to buried facility operators of planned excavations. Whether or not required by statute, these are actions taken by operators of buried facilities to motivate contractors and others to notify of planned excavations so that the surface locations of buried facilities can be marked before excavation begins. When notice centers are established, this also includes the actions of the center to inform contractors and others on using the center to reduce the notification burden.
- Location and surface marking of buried facility locations. The actions of facility operators to identify and mark the surface above buried facilities to aid excavators to not damage marked facilities.
- Protecting marked buried facilities. The actions of excavators to hand dig excavations adjacent to marked buried facilities and to promptly report to the facility operator any damage caused to the facility.

As can be determined from the above, a one-call notification center is but one of several important elements of an excavation damage prevention program. The one call system deals only with the receipt and processing of notifications on planned excavations. Often these centers provide assistance to buried facility operators with educational materials and perform some educational tasks such as providing information on how to use the notification system and the requirements of the state statute.

IMPROVEMENTS IN EXCAVATION DAMAGE PREVENTION

Question. What are NTSB's recommendations for establishing comprehensive, cooperative excavation prevention programs while maintaining the flexibility that NTSB credits as being responsible for the successes seen with one-call?

Answer. The Safety Board believes that the Federal requirements for a damage prevention program should be to establish objectives that must be met by state damage prevention programs, such as by requiring that:

a. excavation notification systems provide coverage statewide;
b. all operators of buried facilities participate in the notice system;
c. all excavators provide notice of planned excavations;
d. operators of buried facilities provide temporary marking of facilities in the area of proposed excavations; excavators protect marked facilities and provide immediate notice of any damage caused to buried facilities; and
e. States monitor compliance with the program requirements and take prompt, effective action when program requirements are not met.

This will allow the States to be innovative in the programs they develop and enable the States to meet needs unique to a particular state due to environment or operating practices. Program operating flexibility over past years has developed effective notice centers operated under contract as well as centers operated by the buried facility operators. Flexibility also has allowed notice centers to seek innovative ways to quickly incorporate new electronic equipment into their operations to make it easier and quicker for excavators to know when buried facility operators had marked their facilities thus allowing work to commence sooner. Flexibility has allowed the States to innovate with diverse monitoring and enforcement programs that provide compliance using far fewer state resources. The Safety Board believes that Federal efforts should concentrate on defining the results to be obtained while providing great latitude for developing the specifics of programs designed to achieve the desired results.

At our September workshop, we will learn if there are other elements necessary for developing effective programs and we will gain more detailed information on each element. We believe that this workshop will provide for both the OPS and the Safety Board the information needed to motivate future improvements in these programs.

QUESTIONS ASKED BY SENATOR PRESSLER AND ANSWERS THERETO BY MR. HALL

Question. The federal government often exempts itself from participation in mandates set forth by Congress. To date, federal agencies, like the Departments of Transportation and Defense, have not been required to participate in one-call systems. Are you aware of third-party damage caused by federal agencies not having to participate fully in one-call systems, either as excavators or as underground utility operators?

Answer. Generally, Federal agencies that operate buried facilities are not required to report accidents that affect their facilities, nor do existing pipeline data bases consistently provide the identity of excavators who damage facilities.

Our review of Research and Special Programs Administration (RSPA) records, studies and other data indicates that excavation damage to buried facilities is a problem to Federal agencies. The RSPA records for 1992 and 1993 show that there were at least five excavation-caused accidents to private pipelines on Federal lands. One of the five accidents was caused by a contractor working for the U.S. Corps of Engineers.

A Federal Aviation Administration (FAA) report, "Cable Cuts: Causes, Impacts, and Preventive Measures," states "that during fiscal years 1989-91, there were 1,444 equipment outages or disruptions caused by third-party cable cuts—an average of one outage every 12 hours." The FAA reported that 10 percent, or 144, cuts occurred on airport property. The consequences of such outages can be devastating. For example, the FAA report cites a May 4, 1991 communications cable cut that disrupted air traffic control operations for five hours and 22 minutes at four of the FAA's 20 major air-traffic control centers.

The Safety Board's investigation of a December 9, 1990, natural gas explosion on Fort Benjamin Harrison in Indiana shows that the Department of Defense operates about 300 gas distribution pipeline systems on Army, Navy, and Air Force installations. It also operates many other utility systems and has allowed private companies to install and operate utility systems on military installations. At the time of the above referenced accident, none of the Department of Defense services met the Federal Minimum Standards for Pipelines (49 CFR 192) which include requirements for conducting excavation damage prevention programs. Since that time, based on our recommendations, the Department of the Army adopted those standards for its buried gas systems. Neither the Department of the Navy nor the Department of the Air Force has responded to our recommendations.

The existing state damage prevention statutes do not protect buried facilities on military and other Federal installations because local and state laws generally are not applicable.

Question. Should the federal government be required to participate fully in one-call systems? Does this legislation we are considering today accomplish this?

Answer. The Safety Board believes that the Federal government should be required to participate fully, as excavators and as buried facility operators, in state damage prevention programs including the one-call notification centers. Many Federal agencies, including the Forest Service, the Strategic Petroleum Reserves, the Federal Housing Administration, the Department of Defense, the Federal Highway Administration, the U.S. Coast Guard, the Federal Aviation Administration, the National Park Service and Bureau of Indian Affairs, and the General Services Administration, could benefit by reducing costs associated with excavation damages and by improving service reliability if they were required to comply with the provisions of the proposed bill.

Questions Asked by Senator Exon and Answers Thereto by Mr. DeLeon

NATIONWIDE TELEPHONE NUMBER FOR ONE-CALL

Question. Is a nationwide, toll-free number necessary for the proper functioning of one-call systems? Is it possible that the system could function just as well with regional or statewide telephone numbers?

Answer. We do not believe that a single number to be used by all excavators nationwide is necessary for the proper functioning of one-call systems. A national number could be used as a backup where an excavator may not know the number for a state in which he or she intends to dig. In that case, the national number could transfer the caller to the appropriate number in the state of intended excavation; or if an excavator is excavating out of state and does not know the correct number, he would use the nationwide number. The national number should not preclude a statewide number, and the system could function just as well with regional or statewide numbers.

GRANTS TO DEVELOP ONE-CALL SYSTEMS

Question. The bill provides for grants to States to develop or upgrade existing one-call systems. Should Congress consider making non-government entities running one-call systems eligible for such grants?

Answer. Yes, at least for a start up period until the system is self-financing through user fees. Most one-call systems are non-profit, quasi-governmental type operations. Their funding comes from fees collected from underground facility operators, but the potential costs of a major expansion of activity may be a strain on their finances. However, once state programs require that all excavators must call and all facility operators must be members, we expect that the additional resources made available will enable the program to be self sustaining.

PARTICIPATION IN ONE-CALL SYSTEMS

Question. Does the Research and Special Programs Administration (RSPA) favor participation of all facility owners and operators in a one-call system? If not, under what circumstances should a facility owner or operator or an excavator be exempted from participation?

Answer. Any exceptions deemed necessary should be narrowly drawn and should be evaluated on whether the exception would significantly increase the probability of damage to underground facilities that could result in economic damage or injury to people, property or the environment. The latest version of the bill includes sufficiently narrow exemptions that we feel the goals of the bill can be met.

ADDITIONAL STEPS TO PREVENT PIPELINE DAMAGE

Question. In addition to this legislation, what other steps might Congress take to aid pipeline damage-prevention programs?

Answer. We believe that this proposed bill adequately addresses all facets of the one-call aspect of damage prevention programs. Other aspects of damage prevention such as actions pipeline operators should take to protect their systems can be addressed through current statutory authority.

MODEL ONE-CALL PROGRAMS

Question. Should the bill's provisions for a model one-call system be changed to require the Secretary of Transportation to decide what a model one-call system should provide?

Answer. No. The bill provides the Secretary the necessary discretion to determine the elements of a model one-call system while providing a sense of Congress as to what elements should be considered. This approach, coupled with the bill's direction that the Secretary consult with the other involved parties in developing the model program, will assure a model program that all parties are more likely to accept.

PENALTIES UNDER THE BILL

Question. Some interested parties have raised the suggestion that penalties In the bill are too detailed. What is RSPA's current thinking on this issue?

Answer. The original version of the bill called for civil penalties of at least $250 and not more than $25,000. The August 10th House version reads "* * * any excavator or facility operator who violates the requirements of the state program shall be liable for an appropriate administrative or civil penalty." The specific penalty provisions for violations of the one-call requirements in the latest version of the bill have been replaced by provisions allowing for state discretion in providing for administrative or civil penalties. States would be authorized to impose heavy penalties when justified by circumstances and lesser penalties for promptly reported incidents. We support this approach as well as the provision for increased penalties for violations which result in serious accidents or damage to an underground facility. We also support decreased penalties where an excavator who has struck an underground facility, promptly reports the strike.

COMPELLING COMPLIANCE WITH THE ACT

Question. Section 4 of the bill allows the Department of Transportation (DOT) to compel, by litigation, compliance. Under what conditions does DOT envision going into court to compel compliance?

Answer. The Departments does not support this provision. The citizen's suit provision in section 4(c) is not likely to motivate the states to continue the growth and improvement in one-call that have been experienced over the last several years. As indicated by those who represented state interests during negotiation on the bill, this provision would only add to already overloaded court dockets without a meaningful benefit.

EFFECTIVENESS OF ONE-CALL: EXCAVATION PREVENTION

Question. Do you share the National Transportation Safety Board's optimistic analysis of the current effectiveness of one-call systems? Please provide the Committee with RSPA's recommendations for improvements to be made and the role to be played by a comprehensive, cohesive excavation damage prevention program.

Answer. We agree with the NTSB's analysis regarding the effectiveness of one-call systems. Coverage of excavators and facility operators must be universal in order to make one-call systems fully effective. In addition, there must be a program in place that guides excavators in describing the intended excavation site and in digging safely around marked facilities. The program must promote prompt and accurate marking of facilities.

MAPPING PROJECTS

Question. There have been concerns about the costs of mapping projects, i.e., mapping pipelines. There are other mapping projects far use throughout the government. Has DOT considered coordinating the mapping projects with other agencies or with the private sector in orders to save resources?

Answer. Yes. RSPA is actively pursuing a collateral working agreement with the U.S. Geological Survey, the Department of Energy, the Federal Energy Regulatory Commission, and the Minerals Management Service for the mapping of natural gas and hazardous liquid pipelines in the United States. Each of these agencies has some information on the location of natural gas and hazardous liquid pipelines. RSPA is also actively pursuing state pipeline information. RSPA intends to use the information already available from these agencies, improve its accuracy, fill in gaps, and return the improved information back to the Federal and state agencies.

QUESTIONS ASKED BY SENATOR PRESSLER AND ANSWERS THERETO BY MR. DELEON

The federal government often exempts itself from participation in mandates set forth by Congress. To date, federal agencies, like the Departments of Transportation and Defense, have not been required to participate in one-call systems.

Question. Are you aware of third-party damage caused by federal agencies not having to participate in one-call systems, either as excavators or as underground utility operators?

Answer. Yes, we are aware of instances of third-party damage caused by federal agencies. Since government owned and operated systems are not subject to DOT's statutory jurisdiction, we have no official record of such damage. We understand that communications lines have been cut three times in recent months at Camp Pendleton. Civilian contractors were responsible twice and a military contractor also cut service. In addition, a major cross-country AT&T fiber optic line was severed at Camp Pendleton.

Question. Should the federal government be required to participate fully in one-call systems? Does the legislation we are considering today accomplish this?

Answer. Yes, we believe that federal government agencies should fully participate in one-call systems. However, the legislation excepts excavators from a one-call system when excavation is undertaken by or for a person on real property owned or leased by that person and the same person operates all underground facilities located at the site of the excavation. Some military bases may fall under this exception.

In answer to your second question, yes, all government underground facility operators and excavators should belong to one-call systems and this is covered by the bill. The problem that will have to be addressed in a realistic manner is that many of the federal facilities do not have good maps or other means of determining where their existing facilities are located. They would be hard pressed to be able to meet present requirements of locating those facilities within the required accuracy of 1-3 feet that is required by the various state laws.

QUESTIONS ASKED BY SENATOR EXON AND ANSWERS THERETO BY MR. DENSLOW

GRANTS TO DEVELOP ONE-CALL SYSTEMS

Question. The bill provides for grants to States to develop or upgrade existing one-call systems. Should Congress make non-government entities operating one-call systems eligible for such grants?

Answer. The Virginia Municipal League feels that the grants to states to develop or upgrade existing One-Call systems should not be extended to non-governmental entities operating One-Call systems. Such an extension would limit the monetary amounts granted to each state.

PARTICIPATION IN ONE-CALL SYSTEMS

Question. Does the National League of Cities (NLC) favor participation of all facility owners and operators in a one-call system? If not, under what circumstances should a facility owner or operator or an excavator be exempted from participation?

Answer. The Virginia Municipal League favors participation of all facility owners and operators in a One-Call system to ensure that the legislation reduces as much risk of future accidents as possible. However, the League strongly urges an amendment to the bill which would remove local governments from the sanctioning provisions of the bill, while still requiring their participation. Local governments, unlike utility companies and excavators, must answer directly to the people. Therefore, no sanction is needed to encourage local governments to take care in their excavating or their operation of utilities.

ADDITIONAL STEPS TO PREVENT PIPELINE DAMAGE

Question. In addition to this legislation, what other steps might Congress take to aid pipeline damage-prevention programs?

Answer. It would be extraordinarily helpful to allow local government officials a greater role in the siting of pipelines as well as the oversight of their operation. More local government involvement is especially important in the area of pipeline regulation as well as the sanctioning process for noncompliance with existing regulations.

FEES FOR OPERATING ONE-CALL SYSTEMS

Question. What are NLC's concerns about the process for arriving at an equitable fee structure for one-call operations?

Answer. The Virginia Municipal League is concerned that the utility industries may end up paying much less than the states for the One-Call systems. The industries are already receiving most of the benefit of the legislation in that it is their utility lines that are being protected from third party damage. To distribute the cost proportionately among the states and utilities seems to be only sweetening the deal for the industry. For this reason, the League feels that the majority of the fees necessary for operating the various One-Call systems must come from the utility companies themselves.

EXEMPTIONS FROM ONE-CALL

Question. Is NLC satisfied that the exemptions in the bill are properly drawn and that the penalty provisions are appropriate?

Answer. The Virginia Municipal League does not support the exemptions currently within the language of the bill. The bill treats government localities as if they were utility operators. Yet, local governments are inherently different. It is the local governments who must answer directly to their constituents when a utility service is cut off due to third party damage. States should be allowed to exempt local government entities from the sanctions imposed by the bill.

QUESTIONS ASKED BY SENATOR PRESSLER AND ANSWERS THERETO BY MS. DENSLOW

Question. The federal government often exempts itself from participation in mandates set forth by Congress. To date, federal agencies, like the Departments of Transportation and Defense, have not been required to participate in one-call systems. Are you aware of third-party damage caused by federal agencies not having to participate in one-call systems, either as excavators or as underground utility operators?

Answer. I am personally not aware of any third party damage caused by federal agencies not having to participate in one-call systems. However, I do not have personal dealings with such agencies and would strongly recommend that you contact them for this information.

Question. Should the federal government be required to participate fully in one-call systems? Does the legislation we are considering today accomplish this?

Answer. The federal government should be required to participate fully in one-call systems. In this manner the intent of the bill is best addressed. It is unclear whether S. 2101 includes the federal government because no definition of person exists. However, in H.R. 4394, a person is defined as including any agency of federal, state or local government. This version of the bill then goes on to define a facility operator and an excavator using this definition of person.

QUESTIONS ASKED BY SENATOR EXON AND ANSWERS THERETO BY MR. MCNIEL

Enclosed are my responses to Senator Exon's and Senator Pressler's written questions on the Comprehensive One Call Act which result from the August 3 hearing.

In addition, I would like to restate an answer to a question asked at the hearing about the importance of a single nationwide telephone number. The concept of a single nationwide number reserved for the purpose of reaching the one-call System operating in the area from which the call is placed has great appeal since it eliminates public confusion regarding what number to use. The single number also forms an excellent focal point for a nationwide public education campaign. Most calls reporting planned excavations are made from the same area in which the excavation site is to be located. For those calls made from outside the area, the simple addition of the appropriate area code would reach the proper system.

Question. The bill provides for grants to States to develop or upgrade existing one-call systems. Should Congress make non-government entities operating one-call systems eligible for such grants?

Answer. My understanding is that the grant program in the legislation primarily is to provide an incentive for states to improve one-call systems which will currently not meet the criteria which are to be developed by DOT in their model program. Once a state has determined that it will legislate to create or improve its one-call system, the state should be able to apply to DOT for a grant. States should have the flexibility to make this grant money available to the non-government entitles operating one-call systems in order to improve their programs and, most important, to enhance public information about the programs. These grants should not continue indefinitely. One-call systems are set up to pay for themselves.

Question. Do AGA and INGAA favor participation of all facility owners and operators in a one-call system? If not, under what circumstances should a facility owner of operator or excavator be exempted from participation?

Answer. AGA and INGAA want all operators of underground facilities that are at risk of being struck by outside excavators included in the program. All excavators undertaking activities that could result in the striking of underground facilities should also be required to call the one-call system. There are circumstances where the facility operator has the exclusive facilities and essentially exclusive access to the land and would be the excavator or oversee a subcontracted excavator. In these circumstances, a reasonable argument can be made for exclusion from participation in the program. However, on public or residential land where there are a number of underground facilities, all excavators and operators of underground facilities should be required to participate.

Question. In addition to this legislation, what other steps might Congress take to aid pipeline damage-prevention programs?

Answer. Congress is taking additional steps to improve pipeline safety by almost doubling the budget of the Office of Pipeline Safety in DOT for the upcoming year. Congress should continue oversight to be assured that these additional monies are spent wisely. Members of Congress should also encourage their states to improve one-call systems after passage of this legislation. Congress should also work with DOT and the pipeline industry to continue research efforts to improve technology that will assist in improving pipeline safety. However, as the Pipeline Safety legislation is reauthorized in the next Congress, Congress should resist requiring more prescriptive, specific technologies or actions by DOT and the industry. Congress should instead encourage DOT and the pipeline industry to establish criteria to determine and manage the particular risks facing each pipeline or distribution company to assure each company has the flexibility to manage those risks most important to its system.

Question. What is the usual way in which one-call programs are funded, and will that method continue to work with the development of a comprehensive one-call system?

Answer. Most one-call systems are funded by the underground facilities operators participating in the program. In some states, excavators also pay a fee into the system. Underground facility operators usually pay per telephone call they receive to mark their lines though sometimes fees are levied on a mileage basis. The fee is different from state to state and is usually negotiated between the underground facility operators and the one-call center operator. This system should not change as a result of passage of this legislation.

Question. How can one-call programs be tailored so that excavators who strike facilities are not afraid to report the damage, thus leaving the facility open to further damage?

Answer. The major focus of one-call should be to prevent damage and encourage reporting of a line that has been struck. States should have the flexibility to impose appropriate administrative or civil penalties and be able to reduce these penalties if a violation is promptly reported. H.R. 4394 as reported by the House Energy and Commerce Committee provides this flexibility.

QUESTIONS ASKED BY SENATOR PRESSLER AND ANSWERS THERETO BY MR. MCNIEL

Question. The Federal Government often exempts itself from participation in mandates set forth by Congress. To date, Federal agencies, like the Departments of Transportation and Defense, have not been required to participate in one-call systems. Are you aware of third-party damage caused by federal agencies not having to participate in one-call systems, either as excavators or as underground utility operators?

Answer. There have been occasions where a federal or state agency or an excavator under contract to the agency has struck a pipeline. Some state programs currently exempt state and/or federal agencies. This legislation includes federal and state agencies, both as excavators and as facility operators, to participate in one-call. We would also like federal agencies to require, as part of their criteria for their excavation contractors or subcontractors, that these parties use the one-call system. They should also report to the affected operator an incident where an underground facility has been struck.

Question. Should the federal government be required to participate fully in one-call systems? Does the legislation we are considering today accomplish this?

Answer. The legislation reported by the House Energy and Commerce Committee does not exempt actions by the federal government. Any federal government excavator or any excavator under contract to the federal government should be required to call the relevant one-call system. The one possible exception to this requirement might be excavation on government property where access is under exclusive control of the government entity and no underground facilities exist except for those owned by that government entity. An example of this circumstance might be a highly secure defense facility but would not include a military base.

Questions Asked by Senator Exon and Answers Thereto by Mr. Boswell

GRANTS TO DEVELOP ONE-CALL SYSTEMS

Question. The bill provides for grants to States to develop or upgrade existing one-call systems. Should Congress make non-government entities operating one-call systems eligible for such grants?

Answer. Grants should be made available to States or non-profit organizations providing one call service to a State. The conditions should be for establishing or enhancing the direct operation of a one call service, i.e., contractor awareness improvement of the communications service or equipment including a required long tape recorder. Mapping or call screening program to improve service to its members.

PARTICIPATION IN ONE-CALL SYSTEMS

Question. Does One-Call Systems International (OCI) favor participation of all facility owners and operators in a one call system? If not, under what circumstances should a facility owner or operator or an excavator be exempted from participation?

Answer. Yes, it is most important that all owners be involved. It's only then that it can be a true "One Call System". The callers efficiency, the system efficiency and the safety of all concerned increase proportional to the amount of participation in the program.

ADDITIONAL STEPS TO PREVENT PIPELINE DAMAGE

Question. In addition to this legislation, what other steps might Congress take to aid pipeline damage-prevention programs?

Answer. Congress can take steps to get all contractor/excavator related parties to educate and provide call before you dig awareness via associations, unions, all federal contracts, equipment manufacturers, rental agencies and other industry specific entities including Soil Conservation, Corps of Engineers, Bureau of Mines, DOD, FAA, etc.

COSTS OF A ONE-CALL SYSTEM

Question. As the representative of one-call system operators, please provide the Committee with some indication of what it costs to run a one-call system?

Answer. The average cost of a large state with statewide coverage is about 3 million dollars annually. PA, OH, MI, GA, VA, IL, IN, AZ, KY, TN, NY, NE are examples. Most, if not all, centers have a discount plan for government entities.

Question. What percentage of the cost of a one-call system is the responsibility of the States?

Answer. With the exception of those states that have enforcement programs the states actual cost of participation is in the ten's of thousands. If they use the information in their permitting program in almost all states there would be a surplus not a cost.

There is a cost associated with enforcement, but other than Connecticut, whose figures have been reported to the House staff each of those states has reports on their programs.

QUESTIONS ASKED BY SENATOR PRESSLER AND ANSWERS THERETO BY MR. BOSWELL

Question. The Federal Government often exempts itself from participation in mandates set forth by Congress. To date, Federal agencies, like the Departments of Transportation and Defense, have not been required to participate in one-call systems. Are you aware of third-party damage caused by federal agencies not having to participate in one-call systems, either as excavators or as underground utility operators?

Answer. I am aware of Soil Conservation or their contractors having accidents; 5 killed including 2 Soil Conservationists in Hudson, Iowa.

Question. Should the federal government be required to participate fully in one-call systems? Does the legislation we are considering today accomplish this?

Answer. As an example, forest fires can be fueled by excavation accidents.

QUESTIONS ASKED BY SENATOR EXON AND ANSWERS THERETO BY MR. THOMAS

Question. The bill provides for grants to States to develop or upgrade existing one-call systems. Should Congress make non-government entities operating one-call systems eligible for such grants?

Answer. The grants are intended to encourage States to develop or upgrade existing programs. In addition to supplying guidelines for development of One Call systems, the most effective existing programs provide for State involvement in oversight and enforcement. Such involvement should be encouraged. As funds are limited, to distribute them to non-government entities as well would hamper the purpose. One-Call systems should be paid for by the users—facility owners, operators and excavators.

Question. Does the Association of Oil Pipe Lines (AOPL) favor participation of all facility owners and operators in a one-call system? If not, under what circumstances should a facility owner or operator or an excavator be exempted from participation?

Answer. The AOPL supports broad-based participation by both excavators and owners or operators of underground facilities. it should be noted that participation by the owner or operator of the underground facility should be (and currently is) in the alternative, since participation by the owner when the operator is already a participant is unnecessarily duplicative. However, participation by federal, state, and local government entities and public utilities, as well as private underground facility owners/operators and excavators is critical to a comprehensive and effective program.

Question. In addition to this legislation, what other steps might Congress take to aid pipeline damage prevention programs?

Answer. Support for education programs aimed at local zoning and siting authorities would be helpful. Underground facilities once in rural areas often have been encroached upon by growing residential and industrial areas. Many incidents caused by third-party damage arise due to development too close to underground facilities.

Question. With respect to the issue of "gathering lines" and their possible exemption from one-call, please provide the Committee with an indication of the number of such lines and whether their exemption would be a serious flaw in a one-call system?

Answer. As the association for interstate oil pipelines, AOPL does not have information concerning the number of gathering lines. Such lines are generally in rural areas on property that is owned or leased by the gathering line owner or operator. We understand that language is being considered which would exempt those gathering lines which constitute little threat to the public safety or the environment.

Question. Is the development of a model one-call system better served by requiring the Secretary of Transportation to institute a rulemaking rather than mandate it in this legislation?

Answer. Yes. The detailed crafting of a model One-Call system is better accomplished through a rulemaking where there is ample time to receive and consider numerous recommendations from interested parties.

QUESTIONS ASKED BY SENATOR PRESSLER AND ANSWERS THERETO BY MR. THOMAS

Question. The Federal Government often exempts itself from participation in mandates set forth by Congress. To date, Federal agencies, like the Departments of Transportation and Defense, have not been required to participate in one-call systems. Are you aware of third-party damage caused by federal agencies not having to participate in one-call systems, either as excavators or as underground utility operators?

Answer. While we do not have statistics concerning third-party damage caused by federal agencies (perhaps the Department of Transportation has data), contractors for the federal government have certainly been involved in pipeline incidents.

Question. Should the federal government be required to participate fully in one-call systems? Does the legislation we are considering today accomplish this?

Answer. The federal government is both a major owner of underground facilities and an important excavator. It too should be, and is, included in this legislation.

QUESTIONS ASKED BY SENATOR EXON AND ANSWERS THERETO BY MR. DEVANEY

GRANTS TO DEVELOP ONE-CALL SYSTEMS

Question. The bill provides for grants to States to develop or upgrade existing one-call Systems. Should Congress make non-government entities operating one-call systems eligible for such grants?

Answer. Grants should be made available to not-for-profit organizations to establish or enhance the direct operations of a one-call service.

PARTICIPATION IN ONE-CALL SYSTEMS

Question. Does the American Public Works Association (APWA) favor participation of all facility owners and operators in a one-call system? If not, under what circumstances should a facility owner or operator or an excavator be exempted from participation?

Answer. Our association favors participation of all facility owners and operators in a one-call system.

In a statement reconfirmed in 1991 by its membership; It is an American Public Works Association (APWA) policy that encourages (1) each chapter to maintain effective liaison with the one-call center or centers serving its area, and (2) all APWA members and organizations employing the APWA members that own or operate underground lines or facilities should subscribe to the one-call center serving their area.

ADDITIONAL STEPS TO PREVENT PIPELINE DAMAGE

Question. In addition to this legislation, what other steps might Congress take to aid pipeline damage-prevention programs?

Answer. Congress can take steps in insuring that Federal Acquisition Regulations include specifications utilizing, procedural, performance and qualification standards for damage prevention. Most contracts have some reference to damage, although enforcement has rarely been pursued.

UNDERGROUND FACILITY MAPPING AND MARKING

Question. The Committee understands that owners and operators of underground facilities are often faced with using inadequate records maintained by the facilities for which they are responsible. What can Congress do in improve facilities' mapping and record keeping systems?

Answer. A major challenge of the future will be to enhance the accessibility, communication, and use of spatially referenced data to improve facilities record keeping. The application of Automated Mapping, Facilities Management and Geographic Information Systems (AM/FM/GIS) coupled with Global Positioning Satellite Systems (GPS) can provide solutions for improved operations and management of public works and utility facilities. Investment in the nation's infrastructure should include the use of this technology.

The use of Subsurface Utility Engineering on highway projects has been supported by the Federal Highway Administration. Congress could encourage States to use this service as a key component for accurately identifying conflicts and for the improvement of records.

Question. Once underground facilities are located and marked what technology is available to ensure that the facility will retain this mark so the public always is aware of the existence of the underground facility?

Answer. Permanent utility line maker signs as well as permanent makers placed in or on curbs, pavement or concrete are available. While in rural areas and for certain facilities they are very appropriate or mandatory, in a more urban area such as the city of Boston where as many as 30 accepted, utilities can be identified, markings of all utilities would not be excepted.

I believe, that the general public does not want to see an outline of the invisible support system that serves them.

QUESTIONS ASKED BY SENATOR PRESSLER AND ANSWERS THERETO BY MR. DEVANEY

Question. The federal government often exempts itself from participation to mandates set forth by Congress. To date, Federal agencies, like the Departments of Transportation and Defense, have not been required to participate in one-call systems. Are you aware of third-party damage caused by federal agencies not having to participate in one-call systems, either as excavators or as underground utility operators?

Answer. To answer your question concerning damage caused by federal agencies, I requested a utility from the western region of the United States to pull a summary of facilities damaged by government agencies. I have provided the listing as a two page attachment to this letter. The search time period was for the past 15 months and is based on claims that the utility has or will be billing these agencies. This utility belongs to one-call systems and had a verification or location been requested from a center they would have responded.

While none of the cases resulted in catastrophic damage, service distributions to the public and costs to the agency and utility could have been avoided.

Question. Should the Federal government be required to participate fully in one-call systems?

Answer. Yes, the Federal government has a responsibility as an excavator or as a facility owner to protect its own facilities and those belonging to others.

Question. Does the legislation we are considering today accomplish this?

Answer. If mandatory participation, without exception can be maintained in The Comprehensive One-Call Notification Act of 1994, it will accomplish this requirement.

ATTACHMENT 1

Outside Plant Facilities Damaged by Government Agencies

State	Name	City	Date
Arizona	BIA	Camp Verde	03–02–94
Colorado:			
	U.S. Dept. Interior	Kremmling	09–17–92
	U.S. Air Force	Aurora	07–13–94
	U.S. Forest Service	Durango	08–11–93
	U.S. Forest Service	Durango	06–12–94
	U.S. Forest Service	Hot Sulphur Springs	08–24–94
	U.S. Forest Service	Steamboat	09–27–93
	U.S. Government	Colorado Springs	09–29–93
Indiana:			
	BIA	Blackfoot	07–22–94
	BIA	Pocatello N	09–08–92
	BLM	Malad	06–16–92
	U.S. Forest Service	Idaho City	05–06–94
	U.S. Forest Service	Island Park	07–11–94
Iowa	None this search.		
Minnesota	None this search.		
Montana:			
	BIA	Billings	06–07–93
	BIA	Billings	11–06–92
	BIA	Hardin	06–06–94
	BIA	Harlem	09–10–92
	BIA	Harlem	09–02–92
	BIA	Harlem	09–03–92
	BIA	Harlem	07–25–93
	BIA	Poplar	06–28–93
	BIA	Lodgegrass	09–08–93
	BIA	Lodgegrass	09–27–93
	BIA	Riverton	08–25–94
Nebraska	None this search.		
New Mexico:			
	BIA	Fort Wingate	06–28–94
	BIA	Fort Wingate	10–14–91
	BIA	Farmington	03–15–93
	BIA	Farmington W	12–10–93

Outside Plant Facilities Damaged by Government Agencies—Continued

State	Name	City	Date
	BIA	Farmington	04–24–93
	BIA	Farmington	04–12–94
	Bureau of Reclamation	Socorro	06–23–93
North Dakota	None this search.		
Oregon:			
	BIA	Warm Springs	09–14–93
	BLM	Sutherlin	07–29–94
	BLM	Rogue River	09–01–93
	Bureau of Reclamation	Pendleton	03–30–94
	U.S. Forest Service	Kooskia	10–27–93
	U.S. Forest Service	Culp Creek	03–29–93
	U.S. Forest Service	Gilchrist	07–09–93
	U.S. Forest Service	Baker	06–30–94
South Dakota	None this search.		
Utah:			
	U.S. Forest Service	Panguitch	08–15–93
	U.S. Government	Bicknell	11–19–93
Washington:			
	BIA	Wellpinit	02–03–93
	BIA	Nespelem	11–10–93
	BIA	Springdale	08–05–93
Wyoming:			
	U.S. Forest Service	Jackson	06–20–94
	U.S. Government Navy	Cheyenne	09–12–93

○

APEX
TP
9780160465871
76901945 – 65